Copyright Page

Title: Designing & Building Space Colonies-
A Blueprint for the Future

Subtitle: A Blueprint for the Future

Revised March 2023

The Living in Space Series Book Three

Copyrighted By Martin K. Ettington 2023

All Rights Reserved

ISBN: 9781099010088

Printed in the United States of America 2023

Designing & Building Space Colonies-A Blueprint for the Future

One of the greatest adventures in the future of humanity will be to construct, work, and live in space based structures.

In this book we look at the history of ideas for living in space, proposed space colony designs, and technology.

This book is updated for 2023 with lots of new chapters and information which will all affect the building of space colonies.

The details of current life support technology on the International Space Station is reviewed, and what technologies will be required for development of large scale space colonies.

What other things in terms of financing and materials availability will be also needed?

And who are the innovators providing new space technologies which will be used going forward to build space infrastructure.

Finally, we conclude with some recommendations to get us ready to build these colonies.

Designing & Building Space Colonies-A Blueprint for the Future

Other books by Martin K. Ettington

Spiritual and Metaphysics Books:
Prophecy: A History and How to Guide
God Like Powers and Abilities*
Enlightenment for Newbies
Removing Illusions to Find True Happiness
Using the Scientific Method to Study the Paranormal
A Compendium of Metaphysics and How to Guides (Six books together in one volume)
Love from the Heart
The Enlightenment Experience
Learn Your Soul's Purpose
Pursuing Enlightenment
A Modern Man's Search for Truth
Use Intuition and Prophecy to Improve Your Life
The Handbook of Spiritual and Energy Healing
Pure Spirituality and God
Memories Before Birth and Reincarnation
Paranormal Abilities and the Yoga Sutras of Patanjali
Mystical and Magical Societies and Practitioners
Important Prophecies of the Future
All about Shapeshifting

Longevity & Immortality:
Physical Immortality: A History and How to Guide
The Commentaries of Living Immortals

Records of Extremely Long Lived Persons
Enlightenment and Immortality
Longevity Improvements from Science
The 10 Principles of Personal Longevity
Telomeres & Longevity
The Diets and Lifestyles of the World's Oldest Peoples
The Longevity Six Books Bundle
Long Lived Plants and Animals

A Guide to Longevity Foods, Diets, and Supplements

Science Fiction:
Out of This Universe
The Immortals of the Interstellar Colony
The Mystic Soldier
The Immortality Sci Fi Bundle
Visiting Many Universes
The History of Science Fiction and Fantasy

The God Like Powers Series:
Human Invisibility
Invulnerability and Shielding
Teleportation
Psychokinesis
Our Energy Body, Auras, and Thoughtforms
The God Like Powers Series—Volume 1 Compilation

The Yoga Discovery Series:
Yoga-An Ancient Art Form
Hatha Yoga-Helping you Live Better
Raja Yoga-Through the Ages
The Yoga Discovery Package

Business & Coaching Books:
Creating, Paublishing, & Marketing Practitioner Ebooks
Building a Successful Longevity Coaching Business
Why Become a Coach?
The Professional Coaching Success Trilogy
2020-Make Money Writing and Selling Books
The 2020 Handbook of High Paying Work Without a College Degree
The important of Creativity and How to Improve Yours

Self-Improvement
Stress Relief and Methods to do So
The Importance of Creativity and How to Improve Yours
Building Self-Confidence
See the World Clearly

A Trilogy of Self Help Books
A New Paradigm of Truth and Happiness
Building Hope and Wonder Among Chaos
The Importance of Genius In Our World
The Fear of Failure

Science, Technology, and Misc.
Future Predictions By and Engineer & Seer
The Unusual Science & Technology Bundle
Removing Limits On Our Consciousness-And Thinking Outside the Box
Universal Holistic Philosophy
Ball Lightning
Stranger Than Science Stories and Facts
Planet Earth is Conscious
Accept Science Paradigms Which May Be Wrong
Infinity and The Unbounded Universe

Survival
Survival of Humanity Throughout the Ages
33 Incredible True Survival Stories
The Importance of Fire in History and Mythology
How to Survive Anything: From the Wilderness to Man Made Disasters
Building and Stocking a Nuclear Shelter for less than $10,000
The Human Survival Five Books Bundle
Stranger Than Science Facts and Stories
Stranger Than Science Facts and Stories Volume Two
The Microscopic World Inside and Around Us

Legendary Beings
Are Cryptozoological Animals Real or Imaginary?
Fire in History and Mythology
All About Dragons

Sea Serpents and Ocean Monsters
The Legendary Animals Five Books Bundle
The Mythical People of Ireland
Bigfoot Mysteries and Some Answers
About the Little People: Fairies, Elves, Dwarfs and Leprechauns
Strange Stories From National Parks

Ancient History
The Real Atlantis-In the Eye of the Sahara
Ancient & Prehistoric Civilizations
Ancient & Prehistoric Civilizations-Book Two
The History of Antediluvian Giants
The Antediluvian History of Earth
Ancient Underground Cities and Tunnels
Strange Objects Which Should Not Exist
More Out of Place Artifacts
Strange and Ancient Places in the USA
A Theory of Ancient Prehistory And Giant Aliens
The Destruction of Civilization About 10,500 B.C.
A Timeline of Intelligent Life on Earth
A 300 Million Year Old Civilization Existed on Earth
The Encyclopedia of Out of Place Artifacts
Hollow and Inner Earth Stories and Facts
The Underground Wall in Rockwall, Texas

Aliens and Space
Types of UFOs Observed in History
Aliens and Secret Technology
Aliens Are Already Among Us
Designing and Building Space Colonies
Humanity and the Universe

Living in Space
All About Moon Bases
All About Mars Journeys and Settlement
The Space and Aliens Six Books Bundle

Designing & Building Space Colonies-A Blueprint for the Future

The Space Colonies and Space Structures Coloring Book
All About Asteroids
Spaceships, Past, Present, and Future
Astronauts, Cosmonauts, and Other Important Space Flyers
All About Mars Journeys and Settlement
Mining the Asteroid Belt
The New Era of Space Stations
Moon Landings, Bases, and Exploration

Time Travel and Dimensions
Real Time Travel Stories From a Psychic Engineer
The Real Nature of Time: An Analysis of Physics, Prophecy, and Time Travel Experiences
Stories of Parallel Dimensions
We Live in a Malleable Reality-and We Can Change It
The Time, Dimensions, and Quantum Mechanical Bundle
Alternate Dimensions & the Otherworld
The Multiverse: Time and Dimensional Travel Q&As
Infinity and Our Unbounded Universe
Quantum Mechanics, Technology, Consciousness, and the Multiverse

Political and Social
The Empire of the United States: Forged By God's Spirit Through Man
The Suppression of Truth in the United States and the World

The Longevity Training Series

(A transcription of the online Multimedia Longevity Coaching Training Program)

The Personal Longevity Training Series-Book1-Long Lived Persons
The Personal Longevity Training Series-Book2-Your Soul's Purpose
The Personal Longevity Training Series-Book3-Enable Your Life Urge
The Personal Longevity Training Series-Book4-Your Spiritual Connection
The Personal Longevity Training Series-Book5-Having Love in Your Heart
The Personal Longevity Training Series-Book6-Energy Body Health
The Personal Longevity Training Series-Book7-The Science of Longevity
The Personal Longevity Training Series-Book8-Physical Body Health
The Personal Longevity Training Series-Book9-Avoiding Accidents
The Personal Longevity Training Series-Book10-Implementing These Principles

The Personal Longevity Training Series-Books One Thru Ten

These books are all available in digital and printed formats from my website and on Amazon, Barnes & Noble, Apple ITunes, and many other sites

My Books Website is: http://mkettingtonbooks.com

Signup for our Mailing List to get the following:

1) A discount coupon for 25% discount on all books on our site
2) Occasional Notices of new books available
3) Occasional Email on other offerings of ours (Monthly)

If you have any questions about this book or other subjects please contact the Author at:

mke@mkettingtonbooks.com

Designing & Building Space Colonies-A Blueprint for the Future

Table of Contents

1.0 Introduction	1
2.0 Realities and Reasons	5
3.0 The History of Space Habitats	9
4.0 Major Proposed Space Habitats	13
The Rotating Wheel	13
O'Neil Cylinders	14
The Stanford Torus	17
The Bernal Sphere	19
Kalpana One	20
5.0 Fringe Technologies	25
6.0 Space Environmental Issues	27
Radiation Protection	27
Heat and Cold	28
Fresh Water	30
Breathable Atmosphere	34
Waste Elimination	38
Power Systems	39
Food Production	45
Atmosphere	50
Guidance and Control	51
7.0 Financial & Other Incentives	55
Government Vs Free Enterprise	56
Space Tourism	58
Space Manufacturing	61
Mars Settlement	63
8.0 The Deep Space Gateway	65
9.0 New Space Stations and Those Being Planned	67
The Chinese Tiangong Space Station	67
Orbital Reef	69
Starlab	71
Concept for a Huge Chinese Space Station	73
10.0 5-15-Years Bases on the Moon	77
11.0 5-15 Years-Hotels in Orbit	81
12.0 15-40 Years-Mars Colonies	87
13.0 Space Infrastructure Development	91
Reduced launch costs	92

Robotic Construction	94
Three Dimensional Printing	100
Solar Smelting	101
Propulsion Systems	103
Nuclear Propulsion	103
Solar Electric Propulsion	109
Ion Propulsion	112
The EM Drive	113
14.0 Leaders in New Technologies	115
Space X-Rocket History	115
SpaceX's New Starship Rocket	119
Blue Origin's Rockets	123
DreamChaser	127
Artemus Moon Landers	131
Planetary Resources-Asteroid Mining	137
3D Printing companies	139
Relativity Space's New Printed Rocket	141
15.0 Materials for Construction	153
Mining the Moon	154
Asteroid Mining	158
Space Elevators	161
15.0 50-100 years-Habitat Cylinders	163
Habitat Phase 1-Getting Started	163
Habitat Building-The Early Years	171
Redwood Forest-Building the Inside	174
Construction Continues	185
Permanent Residency	191
Living in Redwood Forest for Five Years	194
Bora Bora Two	203
17.0 75-150 years-Asteroid Homes	207
My Secret Hideaway	207
18.0 1000+ years Very Large Structures	215
18.1 Ringworld Structures	216
Establishing Gravity	217
Managing the Sun	218
Impossible Strength?	219
From Worlds to a Ringworld	220
18.2 Dyson Spheres	223
Dyson Spheres, Swarms, and Bubbles	224
Dysonian existence	228

19.0 Long Term Planning	231
20.0 Summary	239
21.0 Bibliography	241
22.0 Index	245

Designing & Building Space Colonies-A Blueprint for the Future

1.0 Introduction

I've always been interested in space travel. Having been born in 1955, I grew up hearing and seeing news about the space race between the United States and the Soviet Union.

Walking with friends to elementary school we used to talk about the latest launch of the Mercury Seven like John Glenn and Scott Carpenter.

Then I remember the first manned landing on the moon in 1969. I was at Boy Scout Camp and managed to watch the landing on a TV in the Scout Store. That evening we all assembled in the dining hall to watch the first moon walk. It was really incredible.
I even had a four foot high detailed model of the full Saturn Five rocket in my bedroom.

Contrary to what some conspiracy theorists say today it was all real. The media was saturated with coverage. There were thousands of pictures of men on the moon and lots of live video.

A Couple years later in 1972 I was staying with my grandparents in Florida and watched the astronauts as they moved around riding their moon rover and collecting rocks. I had to become an astronaut.

I became an engineer in the 1970s when I went to Rensselaer Polytechnic Institute, Troy, New York. Some of our graduates had gone on to become Astronauts themselves.

In 1984 I was working for Hewlett Packard in computer sales and had a chance to become part of the team covering Johnson Space Center and the contractors. This gave me the opportunity to visit all of the buildings and see many aspects of manned spaceflight. I also met several astronauts and got to know one guy pretty well who later became an Astronaut. David Wolf was on the Mir Space Station and did construction on the International Space Station.

Also, sadly I knew Judy Resnick who was killed in the Challenger Explosion in 1986. Saw Reagans Speech to NASA up close too for the memorial service. (In fact you can see me in the crowd on some historical footage of the event.)

Of course I applied to become an astronaut too while living in Houston. But I only had a B.S. degree in engineering, not a doctorate which would have given me a real chance. But I did get my pilots license.

As you can tell, I have always had a big passion about outer space. I must have read several thousand science fiction books in my life-many dealing with space travel.

It is also important to point out that most thinking about space habitats was done back in the 1970s-almost fifty years ago. I know from living through that era that there was an optimism in that time that most of us would live to go into space.

The moon landings had just finished and everyone was thinking about the next steps in space in the nineteen seventies. Nobody back then would have believed that we wouldn't go back to the moon even in fifty years.

Take the movie 2001 which came out in 1968 which was right in the middle of the Apollo moon landing plans. One of the scenes shows a space shuttle with a Pan Am Logo going to an orbital space station. (Pan Am was a major international twentieth century airline) This told people that it was only a matter of time until everyone would be going to visit space stations.

Now we know that the last fifty years seen a lack of interest in space and confusion about planning for the future.

It is man's natural destiny to work and live in space. It's just the lack of vision which is the reason why we haven't visited the moon again almost fifty years after the first moon landings.

Having recently started writing science fiction in addition to my other books, I've dealt with habitats in space too. It is an exciting possibility for the future of humanity.

*****

One of the biggest sources of my research is the online library at the National Space Society. It includes several of the classic books on space settlements fully online.

(http://www.nss.org/settlement/library.html)

I haven't seen anywhere a good overview of the whole space habitat movement since the original ones were published in the1970s. Sure, there are some new technical papers, and lots of books by ex-Astronauts, but no recent overviews of this whole area of space habitat conceptualization.

So one of the main goals for this book is to provide an updated overview of space habitat technology given what we have already accomplished with the

International Space Station and the directions of research into applicable space technologies today.

I've also included some sections of a couple of my science fiction books which cover building and living in a space habitat to give the reader an experience of what living and working in Space might look like. This is my own effort to get people excited again about building and living in Space Habitats.

I will consider myself successful if after reading this book you get enthusiastic again about what we can accomplish in space.

2.0 Realities and Reasons

Space travel is also expensive. This is the biggest limitation imposed on working and living in space. We have the technologies to build more space habitations today, but the cost is prohibitive.

The International Space Station cost about $150 billion dollars to build. This is a huge amount of money. NASA is also budgeting $3-4 billion for continued support through the early 2020s.

So we can propose all types of space structures, but the continued cost of space launches will remain the main limiting factor. Until launch costs can be reduced significantly, it will still be difficult to plan things in space.

What SpaceX is doing with re-usable rockets is a real game changer for spaceflight costs.

The physics of building Space Habitats is well understood. It is the development of the technologies and the funding for such these structures which is the challenge.

There are also a number of technologies we will require to build large space habitats including replicating robots, replicating three dimensional printers, propulsion technologies, and more.

Given all these realities, here are a number of reasons listed in the National Space Society website as to why we should build and live in space habitats:

1) Proximity to Earth. The first orbital settlements may well be built only a few hundred miles from Earth in 'Low Earth Orbit' (LEO). High LEO is far enough out that the settlement won't crash into Earth but low enough for the Earth's van Allen Belts to protect settlers from deadly solar storms. Travel back and forth to Earth should take only a few hours. Visits from relatives and friends will be common, and traveling to Earth for vacation or schooling should be easy. Perhaps more important, bringing supplies, materials, and specialized equipment from Earth to support construction will be relatively easy.

2) Continuous, ample, reliable solar energy. In a high enough orbit there is no night. Solar power is available 24/7 in most high orbits, although in high LEO there is some darkness during each orbit as a structure passes through the Earth's shadow. Most satellites in Earth orbit use solar power today, deploying large solar cell arrays like wings stretching from the craft. The solar arrays for settlements must be huge in order to generate enough power. This power can be generated on separate solar power satellites and beamed to the settlement, much as power beamed from such satellites to Earth can play a major role in solving our energy problems.

3) Weightless construction. Zero-g construction means big settlements can be built with relative ease. On Earth, for example, you could not build a round structure that is several miles high because it would collapse under its own weight, but in zero-gravity it is entirely possible to build such large structures, and in orbit astronauts can move objects weighing many tons by hand. Space settlers will spend almost all of their time inside the settlement because it is

impossible for an unprotected human to survive outside for more than a few seconds. In this situation, obviously, bigger settlements are better. Settlements can be made so large that, even though you are really inside, it feels like the out-of-doors.

4) Weightless recreation. Although space colonies will have 1g at the hull, in the center you will experience weightlessness. If you've ever jumped off a diving board, you've been weightless. It's the feeling you have after jumping and before you hit the water. The difference in an orbital space settlement is that the feeling will last for as long as you like. If you've ever seen videos of astronauts playing in 0g, you know that weightlessness is fun. Acrobatics, sports and dance go to a new level when constraints of gravity are removed.

5) Great views of Earth (and eventually other planets). Space settlement is, at its core, a real estate business. The value of real estate is determined by many things, including "the view." Any space settlement will have a magnificent view of the stars at night. Settlements in Earth orbit will have one of the most stunning views in our solar system: the living, ever-changing Earth.

6) Enormous growth potential. If the single largest asteroid (Ceres) were to be used to build orbital space settlements, the total living area created would be well over a hundred times the land area of the Earth. This is because Ceres is a solid, three dimensional object but orbital space settlements are basically hollow with only air on the inside. Thus, Ceres alone can provide the building materials for uncrowded homes for hundreds of billions of people, at least.

7) Economics. Near-Earth orbital settlements can service Earth's tourist, energy, and materials

markets. Space settlements, wherever they are built, will be very expensive. Supplying Earth with valuable goods and services will be critical to paying for settlement.

In this book we will focus on practical stepping stones to building viable space colonies, and what are the incentives—financial or otherwise to have them constructed.

3.0 The History of Space Habitats

"The Brick Moon" written in 1869 is presented as a journal. It describes the construction and launch into orbit of a sphere, 200 feet in diameter, built of bricks. The device is intended as a navigational aid, but is accidentally launched with people aboard. They survive, and so the story also provides the first known fictional description of a space station.

The modern history of space habitats discussions goes back to the 1970s when Gerard O'Neil published his forward thinking book "The High Frontier":

Gerard O'Neil

In the early 1970s America had proved its leadership in Human Spaceflight but among the nation's youth an anti-technology mindset was growing. Princeton Physicist and Professor Dr. Gerard K. O'Neill, inventor of the revolutionary Colliding-Beam Storage Ring technology that is now the basis of all high energy particle accelerators, asked his students if they could come up with a working Space Colony system to permanently and happily house tens of thousands of regular people. They dug into the challenge.

Soon his small band of students grew to scores of researchers both young and old, all united in the Big Dream of letting real people have a real choice in their futures.

In 1974, Dr. O'Neill put his three-pronged plan of Space Colonization, Space Solar Power and Large Scale Space Construction into easily accessible form with the release of the book The High Frontier. Fourteen years later, The Space Studies Institute, founded by O'Neill, re-released the original text, unchanged except for the doctor's addition of the Appendix "A View from 1988."

In 1977 another great collection of thinking and proposals on "Space Colonies" was edited by Stewart Brand. This book is a compilation of studies and articles on space habitats.

Finally, what most space aficionados think is the best historical book on Space Colonization is "Colonies In Space" by T. A. Heppenheimer 1977.

The chapters in this book include the following:

Chapter 1 - Other Life in Space
Chapter 2 - Our Life in Space
Chapter 3 - Power from Space
Chapter 4 - Hope for the Future
Chapter 5 - First of the Great Ships
Chapter 6 - The Moon-Miners
Chapter 7 - Construction Shack
Chapter 8 - The Highest Home
Chapter 9 - Up on the Farm
Chapter 10 - Ventura Highway Revisited
Chapter 11 - What's to Do on Saturday Night?
Chapter 12 - The Shell of the Torus
Chapter 13 - University of Space
Chapter 14 - The Next Million Years
Chapter 15 - Ring Around the Sun
Chapter 16 - Colonizing the Stars

4.0 Major Proposed Space Habitats

Over the last fifty years or so, many space habitat designs have been proposed and lots of engineers, scientists, and visionaries have worked on these concepts. None of them can be built with today's technologies, but they all offer incredible visions for the future.

The Rotating Wheel

Both scientists and science fiction writers have thought about the concept of a rotating wheel space station since the beginning of the 20th century. Konstantin Tsiolkovsky wrote about using rotation to create an artificial gravity in space in 1903. Herman Potočnik introduced a spinning wheel station with a 30-meter diameter in his Problem der Befahrung des Weltraums (The Problem of Space Travel). He even suggested it be placed in a geostationary orbit.

In the 1950s, Wernher von Braun and Willy Ley, writing in Colliers Magazine, updated the idea, in part as a way to stage spacecraft headed for Mars. They envisioned a rotating wheel with a diameter of 76 meters (250 feet). The 3-deck wheel would revolve at 3 RPM to provide artificial one-third gravity. It was envisaged as having a crew of 80.

In 1959, a NASA committee opined that such a space station was the next logical step after the Mercury program.

Designing & Building Space Colonies-A Blueprint for the Future

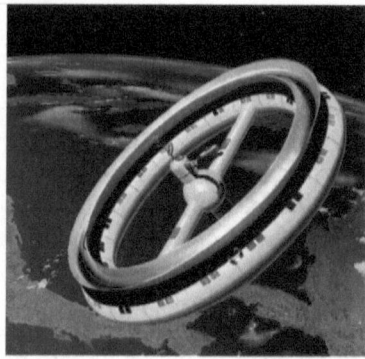

An internal cross section of what the inside of the wheel might look like:

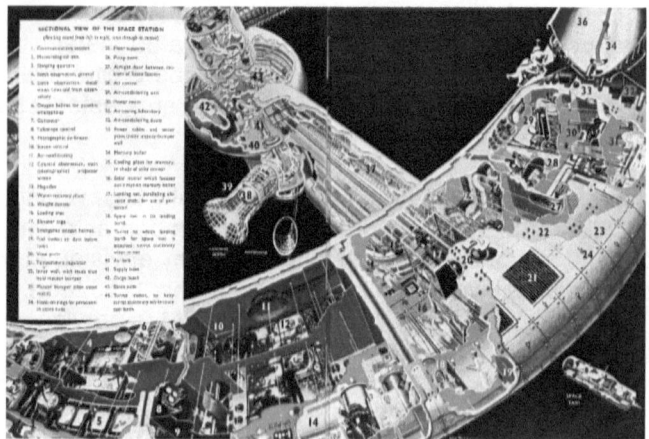

O'Neil Cylinders

The O'Neill cylinder (also called an O'Neill colony) is a space settlement design proposed by American physicist Gerard K. O'Neill in his 1976 book The High Frontier: Human Colonies in Space. O'Neill proposed the colonization of space for the 21st century, using materials extracted from the Moon and later from asteroids.

An O'Neill cylinder would consist of two counter-rotating cylinders. The cylinders would rotate in opposite directions in order to cancel out any gyroscopic effects that would otherwise make it difficult to keep them aimed toward the Sun. Each would be 5 miles (8.0 km) in diameter and 20 miles (32 km) long, connected at each end by a rod via a bearing system. They would rotate so as to provide artificial gravity via centrifugal force on their inner surfaces.

Here is a look at a notional view of the inside of an O'Neil habitat:

Two counter rotating O'Neil Habitats:

A Close-up view of the outside of O'Neil habitats:

The Stanford Torus

The Stanford torus is a proposed NASA design for a space habitat capable of housing 10,000 to 140,000 permanent residents.

The Stanford torus was proposed during the 1975 NASA Summer Study, conducted at American University, and Stanford University, with the purpose of exploring and speculating on designs for future space colonies (Gerard O'Neill later proposed his Island One or Bernal sphere as an alternative to the torus). "Stanford torus" refers only to this particular version of the design, as the concept of a ring-shaped rotating space station was previously proposed by Wernher von Braun and Herman Potočnik.

It consists of a torus, or doughnut-shaped ring, that is 1.8 km in diameter (for the proposed 10,000 person habitat described in the 1975 Summer Study) and rotates once per minute to provide between 0.9g and 1.0g of artificial gravity on the inside of the outer ring via centrifugal force.

Sunlight is provided to the interior of the torus by a system of mirrors. The ring is connected to a hub via a number of "spokes", which serve as conduits for people and materials travelling to and from the hub. Since the hub is at the rotational axis of the station, it experiences the least artificial gravity and is the easiest location for spacecraft to dock. Zero-gravity industry is performed in a non-rotating module attached to the hub's axis.

The interior space of the torus itself is used as living space, and is large enough that a "natural" environment can be simulated; the torus appears similar to a long, narrow, straight glacial valley whose ends curve upward and eventually meet overhead to form a complete circle. The population density is similar to a dense suburb, with part of the ring dedicated to agriculture and part to housing.

The Stanford Torus outside view:

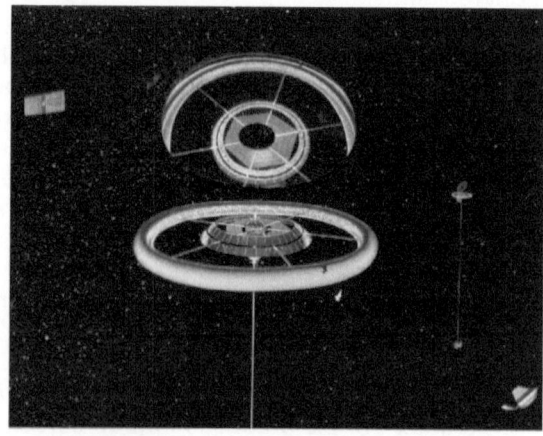

Stanford Torus inside view:

The Bernal Sphere

In a series of studies held at Stanford University in 1975 and 1976 with the purpose of speculating on designs for future space colonies, Dr. Gerard K. O'Neill proposed Island One, a modified Bernal sphere with a diameter of only 500 m (1,600 ft.) rotating at 1.9 RPM to produce a full Earth artificial gravity at the sphere's equator.

The result would be an interior landscape that would resemble a large valley running all the way around the equator of the sphere. Island One would be capable of providing living and recreation space for a population of approximately ten thousand people, with a "Crystal Palace" habitat used for agriculture. Sunlight was to be provided to the interior of the sphere using external mirrors to direct it in through large windows near the poles. The form of a sphere was chosen for its optimum ability to contain air pressure and its optimum mass-efficiency at providing radiation shielding

The interior of a Bernal Sphere:

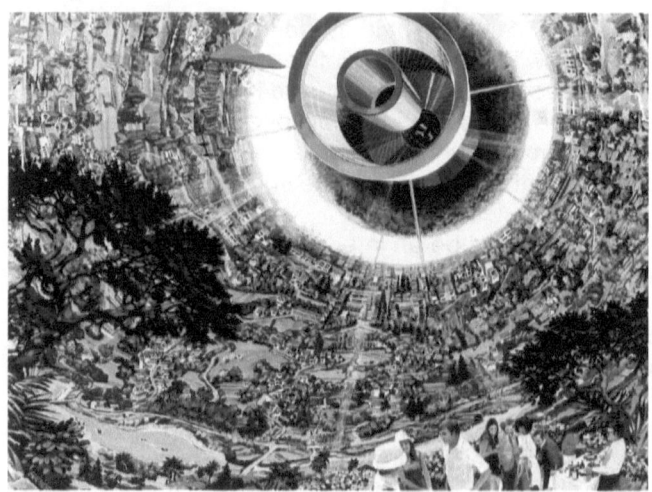

Kalpana One

Kalpana One is intended to be the first, and smallest, of a family of space settlements. The size is determined by the limited rotation rate humans are assumed to tolerate, 2*rpm*. The rotation rate drives the radius to achieve 1*g* pseudo-gravity, and the radius drives the length due to angular moment of inertia requirements. For later, larger settlements in the Kalpana family, the rotation rate may be reduced, increasing the radius and the allowable length.

Kalpana One solves some of the problems found in earlier designs: excessive shielding mass, large appendages, lack of natural sunlight, rotational instability, lack of wobble control, and some catastrophic failure modes. Much is left to be done before a practical space settlement can be fully designed and built. Just as our distant ancestors left the warm oceans and colonized dry land, it is our task to settle the vast, empty reaches of space; thereby ensuring the survival and growth of civilization, humanity, and life itself. Let's get to work.

Here are exterior and interior views:

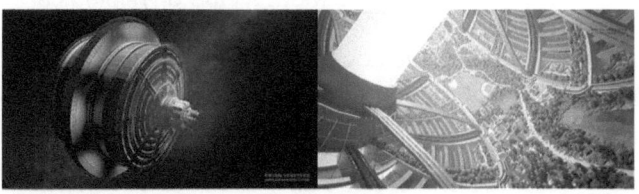

Space Settlement Design Studies

Most space habitat design studies were first done back in the 1970s. Some updates and independent efforts have been produced in the meantime.

One major study was conducted in 1977 by NASA and industry partners. It's called "Space Settlements-A Design Study" NASA SP-413 The full study can be found at:

http://www.nss.org/settlement/nasa/75SummerStudy/Design.html

The design goals are shown here:

> This system is intended to meet a set of specific design goals established to guide the choice of the principal elements of a practicable colony in space. The main goal is to design a permanent community in space that is sufficiently productive to maintain itself, and to exploit actively the environment of space to an extent that permits growth, replication, and the eventual creation of much larger communities. This initial community is to be a first step in an expanding colonization of space.

To effect this main goal, the following subsidiary goals must be met using existing technology and at minimum cost:

- Design a habitat to meet all the physiological requirements of a permanent population and to foster a viable social community.
- Obtain an adequate supply of raw materials and provide the capability to process them. Provide an adequate transport system to carry people, raw materials, and items of trade. Develop commercial activity sufficient to attract capital and to produce goods and services for trade with Earth.

Fortunately, the design study could draw on substantial earlier work. Active interest in space colonization as a practical possibility began in 1969 when Gerard O'Neill and students at Princeton University undertook a detailed assessment of space colonization. They aimed at a model to show the feasibility of a space colony rather than an optimum configuration and they selected as a test case a rotating habitat in satellite orbit around the Earth at the distance of the Moon, using solar energy to sustain a closed ecological system.

They proposed a habitat constructed of processed lunar ore delivered by an electromagnetic accelerator and located at either the Lagrangian point L4 or L5 in order to make delivery of the ore as simple as possible. (The Lagrangian points are described in ch. 2.) The habitat was configured as a 1-km long cylinder with hemispherical end-caps. It was to have an Earth-like internal environment on the inner surface and be supplied with sunlight reflected from mirrors.

Subsequently, the Princeton group suggested that the L5 colony could construct solar power stations from lunar material. They concluded that this would improve the economics of both the satellite solar power stations and the colony itself.

The concept of satellite solar power stations has received increasing attention since its introduction by Peter Glaser in 1968. These ideas were further considered and developed by a conference "Space Manufacturing Facilities" which took place at Princeton University on May 7-9, 1975 and focused more attention on O'Neill's test case.

This report presents a rationale for the design choices of the Ames-Stanford study group and it details how the various parts of the system interrelate and support each other. The next three chapters discuss successively how the properties of space specify the criteria that a successful design must satisfy, what human needs must be met if people are to live in space, and the characteristics of various alternative components of the design. Some readers may wish to skip directly to chapter 5 where the details of the operation of the system are described. Chapter 6 provides a detailed analysis of the sequence of events needed for the colony to be built.

Timetables, manpower requirements, and levels of funding are presented for the construction of the main parts of the overall system. This chapter also looks at long-term benefits from solar power stations in space and some possible ways to structure economics so as to initiate the establishment and growth of many colonies over the long term.

Chapter 7 looks at the future development of colonization of space, and finally chapter 8 discusses why space colonization may be desirable and provides some conclusions and recommendations for further activities and research.

5.0 Fringe Technologies

Most technically knowledgeable fans of Space Colonies will probably shoot me for talking about anti-gravity, but I would be remiss by not mentioning it.

There is much research to suggest that the United States government has had secret projects since the 1940s to develop anti-gravity aerospace and possibly spacecraft. That these craft exist today.

One popular source of this information is in "The Hunt for Zero Point" By Nick Cook published in 2001.

I also wrote my own book on the subject recently titled "Aliens and Secret Technology- A Theory of the Hidden Truth" available on Amazon.com

The availability of Anti-Gravity technology would completely change the design and financing issues for construction of Space Habitats:

1) Space Habitat designs would no longer need centrifugal forces to provide artificial gravity. So designs could be boxes, spheres, or whatever.
2) Materials availabilities for construction would be dramatically increased. It might cost less to haul materials from Earth than from the Moon

3) Labor would cost a lot less and this would reduce the needs to wait for highly automated construction technologies since humans could do a lot of the work—even if living on Earth between shifts.
4) The financial resources needed to build Space Colonies would be significantly reduced since hauling people and materials out of high gravity wells would no longer be an issue.

Anyway, don't worry, since I'm not going to use the availability of Anti-Gravity for any further discussions in this book. This book is all based on our current understanding of Physics and Engineering.

6.0 Space Environmental Issues

There are many environmental issues in space which we will have to address. Many of them have already been solved on a small scale in the International Space Station (ISS)

These systems and approaches will also need to be expanded in a larger construction in space which would support many hundreds and possible thousands of people.

In this chapter we provide information on what ISS does on these issues and what else would need to be done on larger space habitats.

Radiation Protection

On Earth we have a thick atmosphere to protect us from Gamma Rays in general and Solar Flares. In space none of this protection exists and we have to provide it.

On the International Space Station (ISS) in the current ISS, the materials for e.g. the hull are chosen primarily because they are light and strong. Aluminum is common. A few mm of Aluminum blocks most of the radiation you would encounter in low Earth orbit. In the ISS, 95% of the radiation is blocked.

This is enough for low Earth orbit: these orbits are inside the Van Allen Belts, so they are protected from the worst radiation. If we want to go beyond LEO for longer periods, more protection is needed. You could make the hull thicker, but this makes the launch more expensive

Larger Space Habitats:

The larger habitats would have thicker walls and hulls which would be able to take higher radiation levels found outside of Earth's low orbit.

On the Moon or Mars the buildings can be buried to provide more protection. On Space Habitats water can also be put into strategic locations or used for emergency shielding rooms to protect people from radiation.

Heat and Cold

The ISS has a lot of design elements used to maintain and control temperature. (6) Without thermal controls, the temperature of the orbiting Space Station's Sun-facing side would soar to 250 degrees F (121 C), while thermometers on the dark side would plunge to minus 250 degrees F (-157 C). There might be a comfortable spot somewhere in the middle of the Station, but searching for it wouldn't be much fun!

Fortunately for the crew and all the Station's hardware, the ISS is designed and built with thermal balance in mind -- and it is equipped with a thermal control system that keeps the astronauts in their orbiting home cool and comfortable. The first design consideration for thermal control is insulation -- to keep heat in for warmth and to keep it out for cooling.

Here on Earth, environmental heat is transferred in the air primarily by conduction (collisions between individual air molecules) and convection (the circulation or bulk motion of air). "This is why you can insulate your house basically using the air trapped inside your insulation," said Andrew Hong, an engineer and thermal control specialist at NASA's Johnson Space Center. "Air is a poor conductor of heat, and the fibers of home insulation that hold the air still minimize convection."

"In space there is no air for conduction or convection," he added. Space is a radiation-dominated environment. Objects heat up by absorbing sunlight and they cool off by emitting infrared energy, a form of radiation which is invisible to the human eye.

As a result, insulation for the International Space Station doesn't look like the fluffy mat of pink fibers you often find in Earth homes. The Station's insulation is instead a highly-reflective blanket called Multi-Layer Insulation (or MLI) made of Mylar and Dacron.

The reflective silver mesh is aluminized Mylar. The copper-colored material is kapton, a heavier layer that protects the sheets of fragile Mylar, which are usually only 0.3 mil or 3/10000 of an inch thick.

"The Mylar is aluminized so that solar thermal radiation can't get through it," explains Hong. Here on Earth, we use blankets containing aluminized Mylar to wrap people who have been exposed to cold or trauma. Such blankets are especially popular among hunters and campers!

"Layers of Dacron fabric keep the Mylar sheets separated, which prevents heat from being conducted between layers," he continued. "This ensures radiation will be the most dominant heat transfer method through the blanket." Except for its windows, most of the ISS is covered with the radiation-stopping MLI.

There are also active heat tubes and thermal radiation fins to dump excess heat into space.

Larger Space Habitats:

In these larger habitats heat buildup might also be a problem so considerations should be made for having large heat transport pipes and solar radiators outside.

Fresh Water

All water used to be hauled into space by rocket then used up or wasted. The ISS now has a water production system in usage since 2010.

Drinkable water is one of the primary and most important assets for human survival. So when preparing for a journey, whether to sea or to space, planners must take this vital resource into consideration. Stowage space during such voyages always comes at a premium. It is no different for the International Space Station and the resupply vehicles that dock there.

A great example of a solution to minimize size and weight in life support is the recently launched Sabatier system. Originally developed by Nobel Prize-winning French chemist Paul Sabatier in the early 1900s, this process uses a catalyst that reacts with carbon dioxide and hydrogen - both byproducts of current life-support systems onboard the space station - to produce water and methane. This interaction closes the loop in the oxygen and water regeneration cycle. In other words, it provides a way to produce water without the need to transport it from Earth.

The fundamental technology for this particular system has been in development for the past twenty years. The overall schedule for hardware production, however, was under two years. This accelerated timeline was a significant challenge for the complex Sabatier, which contains a furnace, a multistage compressor, and a condenser/phase-separation system. The fact that recycling system feeds for Sabatier were already available on the station helped to simplify some of the design tasks by reducing the unknowns.

According to Jason Crusan, chief technologist for space operations at NASA Headquarters in Washington, the previous development and solid interfaces allowed

NASA to try out a new way of acquiring services for the station with Sabatier. "Being able to demonstrate innovative new methods to acquire technical capabilities is one of the key cornerstones the space station can serve for future missions and approaches to those missions," Crusan explained.

Using developing technologies and productive systems enables the station to squeeze every drop from the resources that must launch from Earth. In addition to improving the efficiency of the station's resupply capabilities, Sabatier also frees up storage space. This helps to maximize the area available for science facilities and engineering equipment. The knowledge gained from such systems also advances the collective understanding of technologies to advance spaceflight and help solve similar problems on Earth.

The Sabatier system has long been a part of the space station plan, but the retirement of NASA's space shuttles elevated the need for new resources to provide water. For a decade, shuttles have provided water for the station as a byproduct of the fuel cells they use to generate electricity. Sabatier supplements the capability of resupply vehicles to provide water to the station, without becoming a sole source for this critical station resource.

Currently in operation on the station, Sabatier is the final piece of the regenerative environmental control and life-support system. This hardware was successfully activated in October 2010 and interacts directly with the Oxygen Generation System, which provides hydrogen, sharing a vent line.

Prior to Sabatier, the Oxygen Generation System vented excess carbon dioxide and hydrogen overboard. Rather than wasting these valuable chemicals, Sabatier enables their reuse to generate additional water for the station. With room and resources at a premium in space, this is a significant contribution to the space station's supply chain.

In addition there is now a degree of water recycling on the ISS. Nature's been recycling water on Earth for eons, and now NASA is set to do the same thing above Earth on the International Space Station. Space shuttle Endeavour carried in two refrigerator-sized racks packed with a distiller and an assortment of filters designed to process astronauts' urine and sweat into clean drinking water.

The station crew depends now on water carried up aboard a space shuttle or cargo rocket. But an operational water recycler is expected to cut that need by 65 percent by producing about 6,000 pounds of potable water each year. That's enough fresh water to allow the station to host six crew members instead of three.

A system that operates on the station also will provide a significant stepping stone to developing even more efficient processes that will support astronauts on the moon or on long-duration voyages into the solar system. Although Russia's space station Mir recycled cosmonaut's sweat, the NASA recycler is the first to be flown in space that intends to cleanse and reuse almost all the water a crew member produces.

The system can recycle about 93 percent of the water it receives, said Bob Bagdigian, the Environmental Control Life Support System project manager at NASA's Marshall Space Flight Center in Huntsville, Ala. The water recycler counts in large part on a distiller that Bagdigian compares to a keg tilted on its side. On Earth, distilling is a simple process of simply boiling water and cooling the steam back into pure water. But without gravity, the contaminants in water never separate from the steam no matter how much heat is used.

"In space, it becomes quite a challenge to distill any liquid in the absence of gravity," Bagdigian said.

So the keg-sized distiller is spun up to produce an artificial gravity field. The contaminants in the urine press against the sides of the drum while the steam gathers in the middle and is pumped to a filter. The filters are not much different from those used on Earth, which means they use charcoal-like materials to pull more unwanted elements from the water. Another process uses chemical compounds that bond with the remaining contaminants so filters can pick them out of the water, too.

"The water that we produce meets or exceeds most municipal water product standards," Bagdigian said. The system has been in different stages of development ever since NASA committed to building a space station in the 1980s. Along the way, individual parts of the system have been flown on space shuttle missions for tests.

The distiller mechanism flew in 2003 and worked just fine in orbit, Bagdigian said. Now the crew of the International Space Station will test the whole apparatus, but they won't drink any at first. Instead, they will take numerous samples and return them to Earth for detailed testing. After the testing is complete, controllers will clear the astronauts to use the fresh water in orbit.

NASA's water filter development has also helped produce filters that are now used in humanitarian efforts to make clean water in areas served only by contaminated sources. The effort to make a crew support system that reduces the need for fresh supplies from Earth includes an oxygen generator that is already installed in NASA's Destiny lab on the space station.

Housed in one rack instead of the two required for the water recycler, the oxygen producer splits the oxygen and hydrogen molecules in water and sends the oxygen into the space station as breathable air. The hydrogen is now dumped overboard. However, another process is under development that will combine the hydrogen with other chemicals that react with each other and produce more water.

While the water recycler in use will work fine for the International Space Station's needs, Bagdigian said work is already under way to make it more efficient so it can be used on long moon exploration missions. "We'll take this system and continue to push its performance and efficiency," Bagdigian said.

Larger Space Habitats:

A large habitat will also need water purification technologies. Some of the need might be alleviated by being able to flow water through vegetation in the habitat to provide some purification.

Breathable Atmosphere

Life support systems on the ISS must not only supply oxygen and remove carbon dioxide from the cabin's atmosphere, but also prevent gases like ammonia and acetone, which people emit in small quantities, from accumulating. Vaporous chemicals from science experiments are a potential hazard, too, if they combine in unforeseen ways with other elements in the air supply.

So, while air in space is undeniably rare, managing it is no small problem for ISS life support engineers. Most people can survive only a couple of minutes without oxygen, and low concentrations of oxygen can cause fatigue and blackouts. To ensure the safety of the crew, the ISS will have redundant supplies of that essential gas.

"The primary source of oxygen will be water electrolysis, followed by O_2 in a pressurized storage tank," said Jay Perry, an aerospace engineer at NASA's Marshall Space

Flight Center working on the Environmental Control and Life Support Systems (ECLSS) project. ECLSS engineers at Marshall, at the Johnson Space Center and elsewhere are developing, improving and testing primary life support systems for the ISS.

Most of the station's oxygen will come from a process called "electrolysis," which uses electricity from the ISS solar panels to split water into hydrogen gas and oxygen gas. Each molecule of water contains two hydrogen atoms and one oxygen atom. Running a current through water causes these atoms to separate and recombine as gaseous hydrogen (H_2) and oxygen (O_2).

The oxygen that people breathe on Earth also comes from the splitting of water, but it's not a mechanical process. Plants, algae, cyanobacteria and phytoplankton all split water molecules as part of photosynthesis -- the process that converts sunlight, carbon dioxide and water into sugars for food. The hydrogen is used for making sugars, and the oxygen is released into the atmosphere.

"Eventually, it would be great if we could use plants to (produce oxygen) for us," said Monsi Roman, chief microbiologist for the ECLSS project at MSFC. "The byproduct of plants doing this for us is food."

However, "the chemical-mechanical systems are much more compact, less labor intensive, and more reliable than a plant-based system," Perry noted. "A plant-based life support system design is presently at the basic research and demonstration stage of maturity and there are a myriad of challenges that must be overcome to make it viable."

Hydrogen that's left over from splitting water will be vented into space, at least at first. NASA engineers have left room in the ECLSS hardware racks for a machine that combines the hydrogen with excess carbon dioxide from the air in a chemical reaction that produces water and methane. The water would help replace the water used to make oxygen, and the methane would be vented to space.

"We're looking to close the loop completely, where everything will be (re)used," Roman said. Various uses for the methane are being considered, including expelling it to help provide the thrust necessary to maintain the Space Station's orbit. At present, "all of the venting that goes overboard is designed to be non-propulsive," Perry said.

The ISS also has large tanks of compressed oxygen mounted on the outside of the airlock module. These were the primary supply of oxygen for the U.S. segment of the ISS until the main life support systems arrived with Node 3 in 2005. After that, the tanks now serve as a backup oxygen supply.

Larger Space Habitats:

This is one area where a large habitat should have a big advantage due to having many more plants and trees. This vegetation will be able to provide most if not all of the oxygen needed in the atmosphere.

Pressure Containment

The main dangers to containing atmosphere on the ISS is for collisions with orbital debris. Even microscopic debris can be dangerous due to the speeds it is traveling of miles per second.

The key assets in Collision Avoidance are:

- The telescopes and radars and satellites in the U.S. Department of Defense Space Surveillance Network that help in detecting, classifying and estimating orbital parameters of space debris
- computers in DoD's Joint Space Operations Center cranking through the large volumes of

data obtained by the surveillance network and identifying dangerous stuff up there
- computers, qualified personnel and procedures at NASA Mission Control Center-Houston
- counterparts of the above in Russia (MCC-M, MSIC)

When Debris Avoidance Maneuvers cannot be performed due to late detection of the threat, risk avoidance procedures force the ISS crew to go for the boats (err, Soyuz vehicles).

Mitigation comes into play after the collision. As discussed elsewhere (Are there any safety procedures in place on the ISS in case of puncture?), if there is time to isolate leaking compartments, the crew may do so. However, repairing the station is considered to be the job of follow-up expeditions. Broadly speaking, there are three other possible solutions to the problem that have not been implemented on the ISS:

- Prevention of debris generation (by responsible design)
- Debris collection
- Active defense (with kinetic interceptors or laser ablation)

Larger Space Habitats:

Extra measures will need to be taken in a large space habitat to make it safe from extreme decompression. This would include self-repairing robots and liquids to reduce large sun window punctures, homes which can maintain air pressure in an emergency, and emergency life support centers strategically placed around the habitat, along with easily available pressure suits people can wear.

Waste Elimination

There are two interfaces used for the toilet. The hose, with the yellow funnel on the end, is used for urination. The commode with the circular aperture is used for defecating.

The astronaut lines themselves up properly and activates airflow that will direct the feces to go in the right direction once it leaves the body.

While liquid waste is indeed recycled into drinking water, solid waste is stored in a tank and that tank is periodically replaced. The full tank is put into a progress module which will burn up in the atmosphere.

Larger Space Habitats:

This should also be much easier on large habitat. The waste can be re-used as fertilizer for crops on a large scale. Also, since the larger habitats will be spun for artificial gravity, normal toilets can be used.

Power Systems

On the ISS there are lots of Solar Arrays to provide power all of the time.

Solar Array Wings

The electrical system of the International Space Station is a critical resource for the ISS because it allows the crew to live comfortably, to safely operate the station, and to perform scientific experiments. The ISS electrical system uses solar cells to directly convert sunlight to electricity. Large numbers of cells are assembled in arrays to produce high power levels.

This method of harnessing solar power is called photovoltaics.

The process of collecting sunlight, converting it to electricity, and managing and distributing this electricity builds up excess heat that can damage spacecraft equipment. This heat must be eliminated for reliable operation of the space station in orbit. The ISS power system uses radiators to dissipate the heat away from the spacecraft. The radiators are shaded from sunlight and aligned toward the cold void of deep space.

Each ISS solar array wing (often abbreviated "SAW") consists of two retractable "blankets" of solar cells with a mast between them. Each wing uses nearly 33,000 solar cells and when fully extended is 35 meters (115 ft.) in length and 12 meters (39 ft.) wide. When retracted, each wing folds into a solar array blanket box just 51 centimeters (20 in) high and 4.57 meters (15.0 ft.) in length. The ISS now has the full complement of eight solar array wings. Altogether, the arrays can generate 84 to 120 kilowatts.

The solar arrays normally track the Sun, with the "alpha gimbal" used as the primary rotation to follow the Sun as the space station moves around the Earth, and the "beta gimbal" used to adjust for the angle of the space station's orbit to the ecliptic. Several different tracking modes are used in operations, ranging from full Sun-tracking, to the drag-reduction mode ("Night glider" and "Sun slicer" modes), to a drag-maximization mode used to lower the altitude. See more details in the article at Night Glider mode.

Batteries

Since the station is often not in direct sunlight, it relies on rechargeable nickel-hydrogen batteries to provide continuous power during the "eclipse" part of the orbit (35 minutes of every 90 minute orbit). The batteries ensure that the station is never without power to sustain life-support systems and experiments. During the sunlit part of the orbit, the batteries are recharged. The nickel-hydrogen batteries have a design life of 6.5 years which means that they must be replaced multiple times during the expected 20-year life of the station. The batteries and the battery charge/discharge units are manufactured by Space Systems/Loral (SS/L), under contract to Boeing. N-H2 batteries on the P6 truss were replaced in 2009 and 2010 with more N-H2 batteries brought by Space Shuttle missions. There are batteries in Trusses P6, S6, P4, and S4.

Since 2017, nickel-hydrogen batteries are being replaced by lithium-ion batteries. On January 6, a multi-hour EVA began the process of converting some of the oldest batteries on the ISS to the new lithium-ion batteries There are a number of differences between the two battery technologies, and one difference is that the lithium-ion batteries can handle twice the charge, so only half as many lithium-ion batteries are needed during replacement. Also, the lithium-ion batteries are smaller than the older nickel-hydrogen batteries. Although they are not quite as long lasting as nickel-hydrogen, they can last long enough to extend the life of ISS.

ISS Electrical Power Distribution

The power management and distribution subsystem operates at a primary bus voltage set to Vmp, the peak power point of the solar arrays. As of 30 December 2005, Vmp was 160 volts DC (direct current). It can change over time as the arrays degrade from ionizing radiation. Microprocessor-controlled switches control the distribution of primary power throughout the station. The battery charge/discharge units (BCDUs) regulate the amount of charge put into the battery. Each BCDU can regulate discharge current from two battery ORUs (Orbital Replacement Unit, a series-connected pack of 38 Ni-H2 cells), and can provide up to 6.6 kW to the Space Station. During insolation, the BCDU provides charge current to the batteries and controls the amount of battery overcharge. Each day, the BCDU and batteries undergo sixteen charge/discharge cycles. The Space Station has 24 BCDUs, each weighing 100 kg.

Larger Space Habitats:

Solar Arrays or nuclear power plants are options for the larger habitats. If the Habitats occupy L5 points or similar locations, they will get almost continuous sunlight. In this case far fewer batteries will be needed, and mainly be used for backup power if the main solar cell energy production goes down. These larger structures will have room to maintain a large power plant. This could be either steam plants operating from focused sunlight energy or a traditional nuclear plant.

Data Architecture/Communications

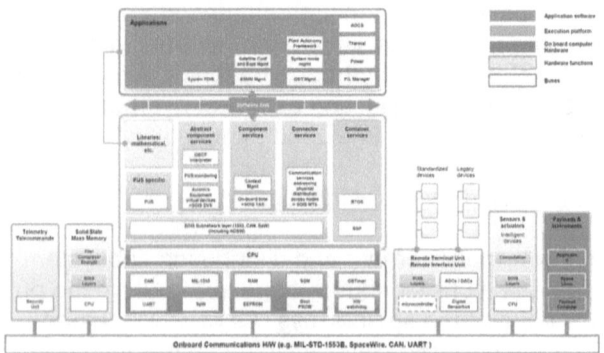

The ISS data architecture and communications system is very complex. I included an architecture diagram above and detailed overview below so that you can see just how much is involved.

On a larger habitat in space imagine that the architecture is that much more complex according to its size and the number of people on it. Fortunately, computing architecture is one area where continuous advances should keep up with the computing needs of an advanced space habitat facility.

Spacecraft Management Unit

On the ISS The On-board Computer (also referred to as Spacecraft Management Unit - SMU or Command & Data Handling Management unit - CDMU) is the central core of the Spacecraft Avionics. The Central Processing Unit (CPU) hosts the Execution Platform SW (composed of RTOS, BSP, SOIS layers, PUS, ...) and the

Application SW. Volatile and Non-volatile Memories, Safe Guard Memories, On Board Timer, Interface controllers and Reconfiguration modules are the other main blocks of a OBC. The figure above shows a functional architecture of the On-Board Data System where all the major functional blocks are indicated with their intercommunication links and their typical redundancy scheme.

Remote Terminal Unit

Remote Terminal Unit (also called Remote Interface Unit-RIU) is a unit that is usually present on medium-large size spacecraft. The RTU offloads the On Board Computer from analogue and discrete digital data acquisition and actuators control tasks.
Platform Solid State Mass Memory

For Earth Observation missions the mass memory for the P/L data may belong to the satellite platform and sometimes, depending on the capacity required, might be included inside the OBC as a single module.
TM/TC

The tele commands, once validated, are multiplexed to the intended addresses. There are two categories of commands: the high priority and the normal commands. The high priority commands (HPC) are sent to the Command Pulse Distribution Unit (CPDU) for immediate execution. The CPDU is either internal to the TC decoder or external and it's implemented in hardware, i.e. no software is involved in the execution of HPCs. The normal commands are sent off to the OBC CPU to be either processed or relayed on the system bus. The Telemetry encoder collects the Telemetry packets from different sources (processing, data storage, essential telemetry, payload), assembles the Telemetry transfer frames and sends them to the TM/TC transceiver to be downloaded to the ground.

Busses

The most common command and control bus used on a spacecraft platform is the MIL-STD-1553B covered by the ECSS-E-ST-50-13C. An alternative to the MIL-STD-1553B is the CAN that ESA and the European Space community is standardizing for space applications. UART serial channels are also used especially to control AOCS sensors.
The Spacewire technology is now being increasingly used for data transfers < 160 Mbit/s and it can combine the command and control function with massive data transfer.

Communication protocols

The space community is asking for a real improvement in the specification and use of communications protocols. Typically, previous developments have harmonized physical interfaces and low level data link protocols but above this level proprietary solutions have been utilized. This has without any doubt increased development and integration costs and limited the possibility of element reuse without expensive modification. In comparison, the commercial market on the ground has systematically pursued the use of multilayer protocol stacks resulting in simple integration and multi-vendor compatibility. This commercial trend is now being adopted for the flight avionics by the development and standardization of protocols above the basic link layer.

Larger Space Habitats:

Data management and communications will of course be critical baseline capabilities on any large habitat. The designs will be similar to those used in the ISS;--just a lot larger and supporting many more people

Food Production

Growing food in space only makes sense. It takes too much energy to ship all your food into space. Also, plants produce oxygen to help with the atmosphere in a habitat. One astronaut on the International Space Station requires approximately 1.8 kilograms of food and packaging per day. For a long-term mission, such as a four-man crew, three year Martian mission, this number can grow to as much as 24,000 pounds.

Due to the cost of resupply and the impracticality of resupplying interplanetary missions the prospect of growing food inflight is incredibly appealing. The existence of a space farm would aid the creation of a sustainable environment, as plants can be used to recycle wastewater, generate oxygen, continuously purify the air and recycle feces on the space station or spaceship. Just 10m² of crops produces 25% of the daily requirements of 1 person, or about 180-210grams of oxygen. This essentially allows the space farm to turn the spaceship into an artificial ecosystem with a hydrological cycle and nutrient recycling.
In addition to maintaining a shelf-life and reducing total mass, the ability to grow food in space would help reduce the vitamin gap in astronaut's diets and provide fresh food with improved taste and texture. Currently,

much of the food supplied to astronauts is heat treated or freeze dried. Both of these methods, for the most part, retain the properties of the food pre-treatment. However, vitamin degradation during storage can occur. A 2009 study noted significant decreases in vitamins A, C and K as well as folic acid and thiamin can occur in as little as one year of storage. A mission to Mars could require food storage for as long as five years, thus a new source of these vitamins would be required.

Supply of foodstuffs to others is likely to be a major part of early off-Earth settlements. Food production is a non-trivial task and is likely to be one of the most labor-intensive, and vital, tasks of early colonists. Among others, NASA is researching how to accomplish space farming.

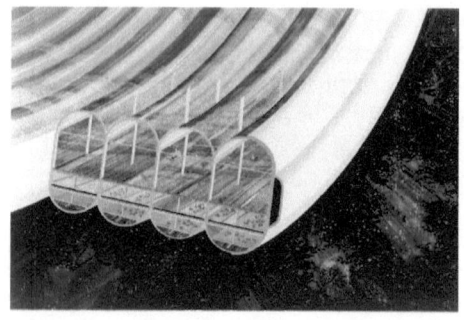

More space food production studies:

ASA plant physiologist Ray Wheeler, Ph.D., and fictional astronaut Mark Watney from the movie "The Martian" have something in common — they are both botanists. But that's where the similarities end. While Watney is a movie character who gets stranded on Mars, Wheeler is the lead for Advanced Life Support Research activities in the Exploration Research and Technology Program at Kennedy Space Center, working on *real* plant research.

"The Martian movie and book conveyed a lot of issues regarding growing food and surviving on a planet far from the Earth," Wheeler said. "It's brought plants back into the equation."

As NASA prepares the Space Launch System rocket and Orion spacecraft for Exploration Mission-1, it's also turning its attention to exploring the possibilities of food crops grown in controlled environments for long-duration missions to deep-space destinations such as Mars.

Wheeler and his colleagues, including plant scientists, have been studying ways to grow safe, fresh food crops efficiently off the Earth. Most recently, astronauts on the International Space Station harvested and ate a variety of red romaine lettuce that they activated and grew in a plant growth system called Veggie.
Wheeler, who has worked at Kennedy since 1988, was among the plant scientists and collaborators who helped get the Veggie unit tested and certified for use on the space station. The plant chamber, developed by Orbitec through a NASA Small Business Innovative Research Program, passed safety reviews and met low power usage and low mass requirements for use on the space station.

Aside from the chamber, the essentials needed for growing food crops, whether on the Earth or another planet, such as Mars, are water, light and soil, along with some kind of nutrients to help them grow.

Potato Crop Studies

A variety of red potatoes called Norland were grown in the Biomass Production Chamber inside Hangar L at Cape Canaveral Air Force Station in Florida during a research study in 1992.

What kind of crops could be grown in space or on another planet? Potatoes, sweet potatoes, wheat and soybeans would all be good according to Wheeler because they provide a lot of carbohydrates, and soybeans are a good source of protein.

Also, potatoes are tubers, which means they store their edible biomass in underground structures. Wheeler said potatoes could produce twice the amount of food as some seed crops when given equivalent light. After salad crops that are now being studied, they are the next category of minimally processed food crops and could be consumed raw.

"You could begin to grow potatoes, wheat and soybeans, things like that, and along with the salad crops, you could provide more of a complete diet," Wheeler said.

Wheeler has spent a lot of time studying different ways to grow potatoes. Most of his studies took place during the late 1980s through the early 2000s inside Hangar L at Cape Canaveral Air Force Station in Florida. The lab was relocated to the Space Life Sciences Laboratory in 2003. A major portion of the labs were then relocated to the Space Station Processing Facility in 2014 to become part of the Exploration Research and Technology Programs Directorate at Kennedy.

Many of the early potato crop studies were done at the University of Wisconsin, where Wheeler worked prior to coming to Kennedy. Plant scientists at Kennedy used these fundamental findings as a starting point for their studies, and in particular, a variety called Norland red potatoes, using a large plant chamber called the Biomass Plant Production Chamber.

The Biomass Production Chamber originally was a hypobaric test chamber used during the Mercury Project. Including its pedestal, the chamber is 28 feet tall. It was later modified to grow plants in the mid-1980s. Air circulation ducts and fans, high pressure sodium lamps, cooling and heating systems, and hydroponic trays and solution tanks were added. The chamber provided a tightly closed atmosphere for plant growth, which simulated what might be encountered in space.

"Providing food is a complex issue," Wheeler said. "We have to think about nutritional issues, what's acceptable and what tastes good. If nobody wants to eat it, that won't work."

Larger Space Habitats:

Food production will be much easier on a large habitat with the room for conventional fields to be planted with crops and even the ability to host meat animals in the large space available due to artificial gravity.

Atmosphere

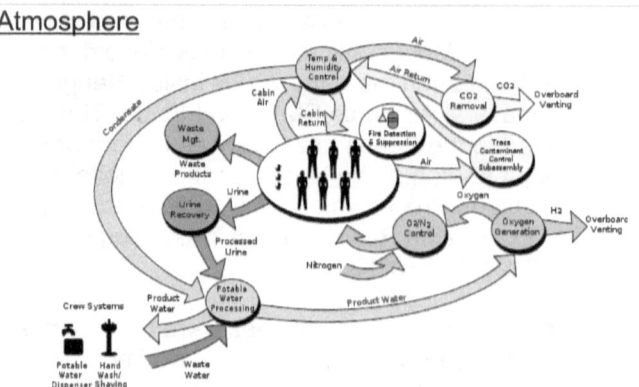

On the ISS the air has to be circulated continuously and everywhere in the station to make sure it is all breathable. This will become an even bigger issue on larger space habitats.

The International Space Station Environmental Control and Life Support System (ECLSS) is a life support system that provides or controls atmospheric pressure, fire detection and suppression, oxygen levels, waste management and water supply. The highest priority for the ECLSS is the ISS atmosphere, but the system also collects, processes, and stores waste and water produced and used by the crew—a process that recycles fluid from the sink, shower, toilet, and condensation from the air.

The Elektron system aboard Zvezda and a similar system in Destiny generate oxygen aboard the station. The crew has a backup option in the form of bottled oxygen and Solid Fuel Oxygen Generation (SFOG) canisters. Carbon dioxide is removed from the air by the Russian Vozdukh system in Zvezda, one Carbon Dioxide Removal Assembly (CDRA) located in the U.S. Lab module, and one CDRA in the U.S. Node 3 module. Other by-products of human metabolism, such as methane from the intestines and ammonia from sweat, are removed by activated charcoal filters or by the Trace Contaminant Control System (TCCS).

Larger Space Habitats:

Atmospheric content should be easy to maintain on a large habitat with the balance between the respiration of vegetation and humans/animals. The main issue will be circulation of the air. Large fans should be supplied to enhance air circulation.

Guidance and Control

Guidance, navigation and control (abbreviated GNC, GN&C, or G&C) is a branch of engineering dealing with the design of systems to control the movement of vehicles, especially, automobiles, ships, aircraft, and spacecraft. In many cases these functions can be performed by trained humans. However, because of the speed of, for example, a rocket's dynamics, human reaction time is too slow to control this movement. Therefore, systems—now almost exclusively digital electronic—are used for such control. Even in cases where humans can perform these functions, it is often the case that GNC systems provide benefits such as alleviating operator work load, smoothing turbulence, fuel savings, etc. In addition, sophisticated applications of GNC enable automatic or remote control.

Guidance refers to the determination of the desired path of travel (the "trajectory") from the vehicle's current location to a designated target, as well as desired changes in velocity, rotation and acceleration for following that path. Navigation refers to the determination, at a given time, of the vehicle's location and velocity (the "state vector") as well as its attitude. Control refers to the manipulation of the forces, by way of steering controls, thrusters, etc., needed to execute guidance commands whilst maintaining vehicle stability.

Keeping the ISS in Place

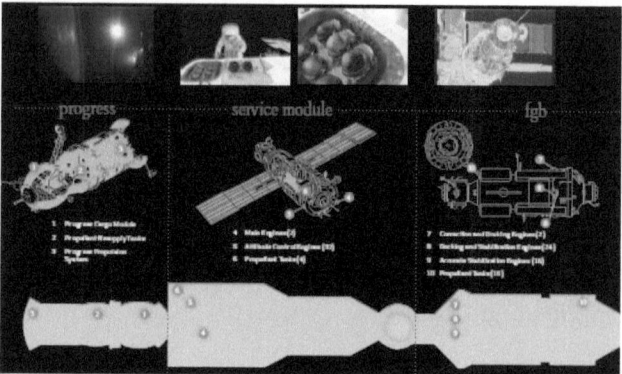

The ISS orbits Earth at an altitude that ranges from 370 to 460 kilometers (230 to 286 miles) and a speed of 28,000 kilometers per hour (17,500 miles per hour). Owing to atmospheric drag, the ISS is constantly slowed. Therefore, the ISS must be re-boosted periodically in order to maintain its altitude. The ISS must sometimes be maneuvered in order to avoid debris in orbit.

Furthermore, the ISS attitude control and maneuvering system can be used to assist in rendezvous and dockings with visiting vehicles, although that capability is not usually required.

Although the ISS typically relies upon large gyrodynes, which utilize electrical power, to control its orientation, when force that is beyond the production capability of the gyrodynes is required, rocket engines provide propulsion for reorientation. (Gryodynes are large fly wheels) Rocket engines are located on the Service Module, as well as on the Progress, Soyuz, and Space Shuttle spacecraft.

The Service Module provides 32 13.3-kilograms force (29.3-pounds force) attitude control engines. The engines are combined into two groups of 16 engines each, taking care of pitch, yaw, and roll control. Each Progress provides 24 engines similar to those on the Service Module. When a Progress is docked at the aft Service Module port, these engines can be used for pitch and yaw control. When the Progress is docked at the Russian Docking Module, the Progress engines can be used for roll control.

Besides being a resupply vehicle, the Progress provides a primary method for reboosting the ISS. Eight 13.3-kilograms force (29.3-pounds force) Progress engines can be used for reboosting. Engines on the Service Module, Soyuz vehicles, and Space Shuttle can also be used. The Progress can also be used to resupply propellants stored in the FGB that are used in the Service Module engines. The ESA ATV and JAXA HTV will also provide propulsion and reboost capability.

Larger Space Habitats:

Keeping the Space Habitats in place will need some large maneuvering thrusters. To reduce propellant needs electric thrusters should be considered.

7.0 Financial & Other Incentives

It doesn't matter that we have the technology if we don't have the money to settle space. The old Mercury Astronauts used to joke about the question of "What makes rockets go up?" The answer was "Funding".

To this point in space history governments such as the United States, Russia, and China have supported and fully paid for manned space flight.

The settlement of outer space has a lot of similarities to the explorations of Europe and their settlement efforts in the fifteenth through seventeenth centuries.

Governments or government funded exploration were first-to find and discover new lands. Then some companies started economic ventures in the newly discovered areas. A good example of this is the Hudson's Bay Company which got British government permission to develop the fur trade in North America from England.

The British East India Company also received permission from the British government to build businesses in India.
In terms of settlements, the Puritans in New England and the settlers of Jamestown were both private ventures to see if profits could be made from settlements in the new world.

Space will be similar. Government pays for the initial exploration while companies will have the incentives to provide space settlement and develop business in space.

Government Vs Free Enterprise

One of the smartest moves NASA made in recent years was to allow private companies to bid on cargo and astronaut transportation to the ISS to focus on further out space exploration. This decision does two things. First it incentivizes private companies to develop and man rate their own rockets and pays them for trips to the ISS. Secondly, it lets NASA get out of standard delivery work and focus its research budget on those things NASA is best at—developing and deploying new space transportation systems.

Here is a picture of the SpaceX Dragon capsule used by Astronauts to get to the ISS. It is a modern design with almost all of the controls on computer displays. This simplification saves a lot of money on reduced wiring.

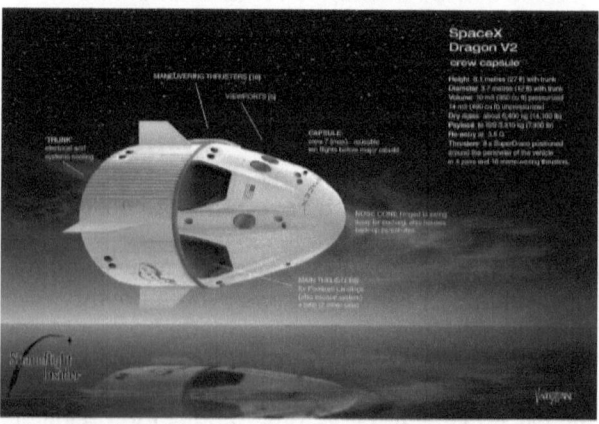

NASA is not a competitive cost efficient organization and should not be doing work which better for more efficient private companies to work on.

Additionally, everything which can be done to get private enterprise involved in space ventures will help everyone in the long run.

Another great book by T. A. Heppenheimer published in 1979 is "Towards Distant Suns" In this book there is a key chapter titled "Paths to Commerce" (3) Here is an excellent diagram showing the paths to commerce in CisLunar space:

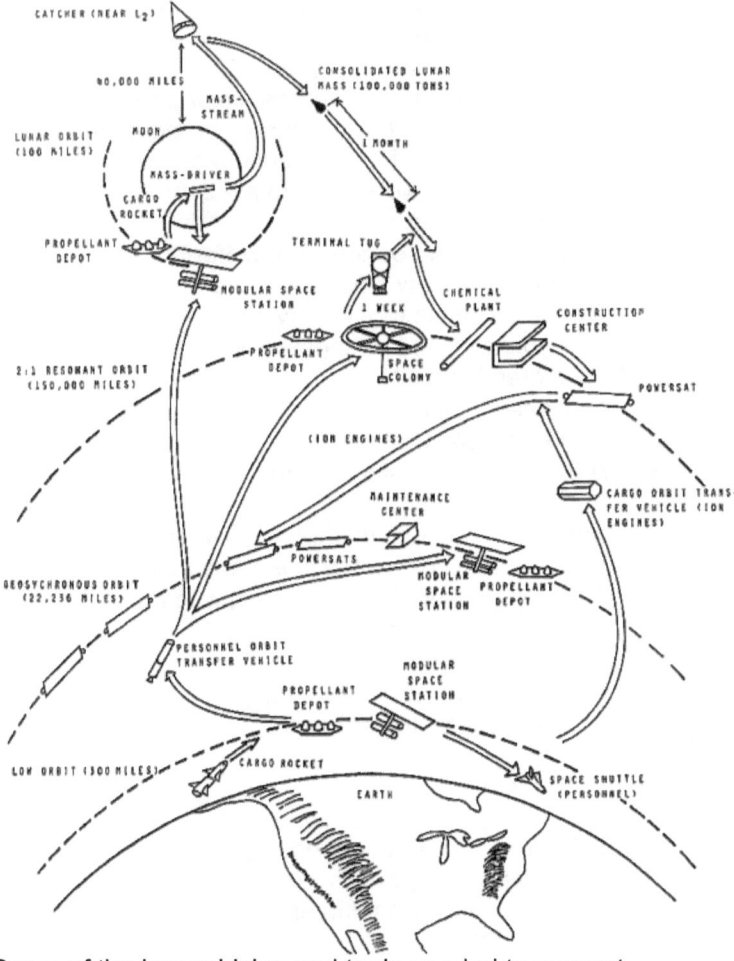

Some of the key vehicles and tools needed to support this commerce and construction infrastructure include:

- Cargo rockets, to carry some five hundred tons of payload from Earth to low orbit.
- An advanced version of the space shuttle to carry seventy-five people to and from low Earth orbit.
- Personnel Orbit Transfer Vehicles to carry people as well as small priority cargoes. These provide connecting service between low Earth orbit, geosynch, the 2:1 orbit, and a low orbit of the Moon.
- Cargo Orbit Transfer Vehicles, solar-powered and with ion drive, to carry large cargoes from low Earth orbit to the 2:1. These may also tow complete powersats from the 2:1 to geosynch on the return leg.
- Space propellant depots, in low Earth orbit, geosynch, low lunar orbit, and at the 2:1. They will be capable of liquefying oxygen and perhaps hydrogen, so that these propellants can be stored indefinitely.
- A maintenance and repair center for the powersats in geosynch.
- The space processing and manufacturing center at the 2:1 orbit.
- Lunar ferry rockets, to shuttle between an orbiting lunar station and the site of the mass-driver.
- The mass-driver and lunar base, with a mine for lunar soil.
- The mass-catcher.
- And, not to be overlooked, the powersats. Most of them will be in geosynch.

Space Tourism

Space Tourism has huge revenue potentials. Some persons have already paid in excess of $20 million dollars each over the last fifteen years to be flown to the ISS to be tourists there.

Recently in 2017 SpaceX CEO Elon Musk said they had two space tourists they were going to launch into orbit around the Moon. How much is he charging each of them?

Here are some of the sub-orbital space tourism proposed and in progress ventures:

Blue Origin is developing the New Shepard reusable suborbital launch system specifically to enable short-duration space tourism.

On October 4, 2004, SpaceShipOne, designed by Burt Rutan of Scaled Composites, won the $10,000,000 X Prize, which was designed to be won by the first private company who could reach and surpass an altitude of 100 km (62 mi) twice within two weeks. The altitude is beyond the Kármán Line, the arbitrarily defined boundary of space. The first flight was flown by Michael Melvill on June 21, 2004, to a height of 100 km (62 mi), making him the first commercial astronaut. The prize-winning flight was flown by Brian Binnie, which reached a height of 112.0 km (69.6 mi), breaking the X-15 record.

Virgin Galactic, headed by Sir Richard Branson's Virgin Group, hopes to be the first to offer regular suborbital spaceflights to paying passengers, aboard a fleet of five SpaceShipTwo-class spaceplanes. The first of these spaceplanes, VSS Enterprise, was intended to commence its first commercial flights in spring 2015, and tickets were on sale at a price of $200,000 (later raised to $250,000). However, the company suffered a considerable setback when the Enterprise broke up over the Mojave Desert during a test flight in October 2014. Over 700 tickets had been sold prior to the accident. A second spaceplane, VSS Unity, has begun testing.

XCOR Aerospace was developing a suborbital vehicle called Lynx until development was halted in May 2016. The Lynx will take off from a runway under rocket power.

Unlike SpaceShipOne and SpaceShipTwo, Lynx will not require a mothership. Lynx is designed for rapid turnaround, which will enable it to fly up to four times per day. Because of this rapid flight rate, Lynx has fewer seats than SpaceShipTwo, carrying only one pilot and one spaceflight participant on each flight. XCOR expect to roll out the first Lynx prototype and begin flight tests in 2015. It was hoped that Lynx would carry paying customers before the end of 2016.

Citizens in Space, formerly the Teacher in Space Project, is a project of the United States Rocket Academy. Citizens in Space combines citizen science with citizen space exploration. The goal is to fly citizen-science experiments and citizen explorers (who travel free) who will act as payload operators on suborbital space missions. By 2012, Citizens in Space had acquired a contract for 10 suborbital flights with XCOR Aerospace and expected to acquire additional flights from XCOR and other suborbital spaceflight providers in the future. In 2012 Citizens in Space reported they had begun training three citizen astronaut candidates and would select seven additional candidates over the next 12 to 14 months.

Space Expedition Corporation was preparing to use the Lynx for "Space Expedition Curaçao", a commercial flight from Hato Airport on Curaçao, and planned to start commercial flights in 2014. The costs were $95,000 each.

Armadillo Aerospace was developing a two-seat vertical takeoff and landing (VTOL) rocket called Hyperion, which will be marketed by Space Adventures. Hyperion uses a capsule similar in shape to the Gemini capsule. The vehicle will use a parachute for descent but will probably use retrorockets for final touchdown, according to remarks made by Armadillo Aerospace at the Next Generation Suborbital Researchers Conference in February 2012. The assets of Armadillo Aerospace were sold to Exos Aerospace and while SARGE is continuing to be developed, it is unclear whether Hyperion is still being developed.

EADS Astrium, a subsidiary of European aerospace giant EADS, announced its space tourism project on June 13, 2007.

The Sierra Nevada Dream Chaser spaceplane is now undergoing developmental testing. It can be used for space tourism or other commercial missions.

Space Manufacturing

One of the main reasons to justify the International Space Station was to experiment and develop space manufacturing techniques. There are some things which can be manufactured in space which are very difficult or impossible on Earth.

The first example of space products sold were microscopic spheres manufactured in space on the Challenger Space Shuttle in 1985:

The Government today announced the sale of the first products made in space, plastic beads so small that 18,000 of them could fit on the head of a pin.

"This material is the first of what we expect will be a long line of products to carry a 'made-in-space' label," said

James M. Beggs, Administrator of the National Aeronautics and Space Administration.

Among other promising space products likely to come along soon, Mr. Beggs said, are pure drugs "to fight age-old diseases," perfect crystals larger than any made on earth and thin films for industrial uses.

Nearly a billion of the tiny spheres produced in the absence of gravity aboard the space shuttle Challenger have been sold to eight companies, the Food and Drug Administration and the University of Utah for use as microscopic yardsticks.

A vial containing about 30 million of the microballs sells for $384, and 29 vials have been sold. The proceeds will be split by the space agency, which produced them, and the National Bureau of Standards, which measured and packaged them.

There are thought to be a number of additional useful products that can potentially be manufactured in space and result in an economic benefit. Research and development is required to determine the best commodities to be produced, and to find efficient production methods. The following products are considered prospective early candidates:

- Growth of protein crystals
- Improved semiconductor wafers
- Micro-encapsulation

As the infrastructure is developed and the cost of assembly drops, some of the manufacturing capacity can be directed toward the development of expanded facilities in space, including larger scale manufacturing plants. These will likely require the use of lunar and asteroid materials, and so follow the development of mining bases.

A picture of protein crystal growth in space:

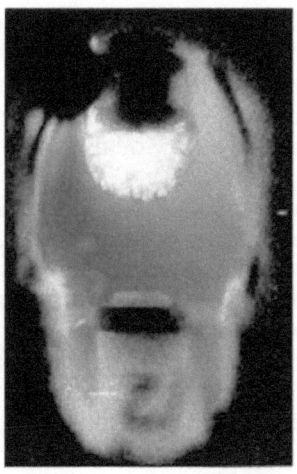

Mars Settlement

It is amazing what people are willing to do to go into space or to Mars. Here is an introduction from the Mars One website on people signing up for a one way paid trip to Mars:

Mars One Astronauts

For anyone not interested to go to Mars, moving permanently to Mars would be the worst kind of punishment. Most people would give an arm and a leg to be allowed to stay on Earth so it is often difficult for them to understand why anyone would want to go.

Yet many people apply for Mars One's mission and these are the people who dream about someday living on Mars. They would give up anything for the opportunity and it is often difficult for them to understand why anyone would not want to go. However, not everyone who wants to fly to Mars is the right type of person to settle on Mars, therefore careful consideration must be taken when considering Mars One's astronauts.

8.0 The Deep Space Gateway

The Deep Space Gateway concept has developed a lot more since 2017 and companies are already building some of the modules. This Lunar Gateway will pioneer deep space habitats and will be an important gateway for the Moon. Systems built for this Gateway may also be used for an actual Mars Manned Mission spaceship.

<u>Lunar Gateway Partners</u>

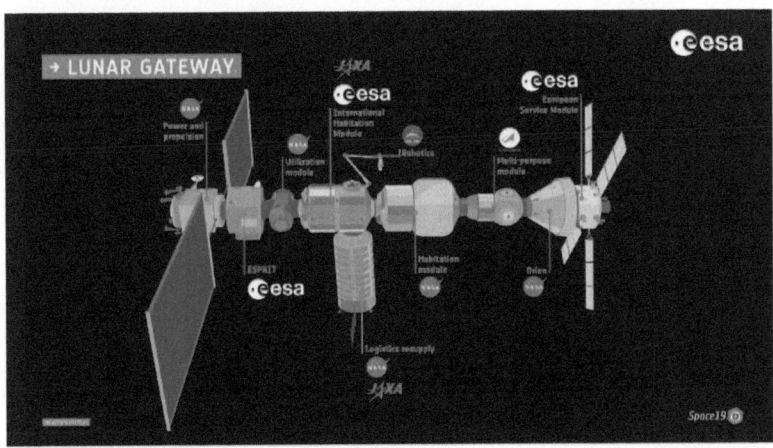

NASA-NASA will build a number of modules including:

- The Power and Propulsion Element
- The US Utilization Module
- The US Habitation Module

Jaxa-The Japanese Space Agency will be helping on the International Habitation Module and the Logistics Resupply Module

Canada-They will provide a robotic arm like they already have for the Space Shuttle and the International Space Station

ESA-The European Space Agency is building several key items. These include the Service Module for the Orion Space Capsule, the International Habitation Module for the Lunar Gateway, and the Esprit Module also for the Lunar Gateway.

Roscosmos-The Russian Space Agency will build the Lunar Gateway Multipurpose Module.

See my book titled "All About Moon Bases" for many more details on the Deep Space Gateway, and the history and plans for new Moon bases.

9.0 New Space Stations and Those Being Planned

Since the first version of this book was written in 2017 the Chinese have launched a space station and new ones are also being planned for later in this decade. So now there is the Tiangong and International Space Stations in Orbit around the Earth.

The Chinese Tiangong Space Station

Tiangong officially the Tiangong space station, is a permanently crewed space station constructed by China and operated by China Manned Space Agency (CMSA) in low Earth orbit between 340 and 450 km (210 and 280 mi) above the surface. It is China's first long-term space station, part of the Tiangong program and the core of the "Third Step" of the China Manned Space Program

(CMS); it has a pressurized volume of 340 m³ (12,000 cu ft), slightly over one third the size of the International Space Station.

The construction of the station is based on the experience gained from its precursors, Tiangong-1 and Tiangong-2. The first module, the Tianhe ("Harmony of the Heavens") core module, was launched on 29 April 2021, followed by multiple crewed and uncrewed missions and two more laboratory cabin modules Wentian ("Quest for the Heavens") launched on 24 July 2022 and Mengtian ("Dreaming of the Heavens") launched on 31 October 2022 The station aims to provide opportunities for space-based experiments, build capacity for scientific and technological innovation

Orbital Reef

Orbital Reef is a planned low Earth orbit (LEO) space station designed by Blue Origin and Sierra Nevada Corporation's Sierra Space for commercial space activities and space tourism uses. Blue Origin has referred to it as a "mixed-use business park" The companies released preliminary plans on 25 October 2021. The station is being designed to support 10 persons in 830 m3 of volume. The station is expected to be operational by 2027.

On 2 December 2021, NASA announced it had selected Blue Origin as one of three companies to develop designs of space stations and other commercial destinations in space. Blue Origin was awarded $130 million. These Space Act Agreements are the first phase of two with which NASA aims to maintain an uninterrupted U.S. presence in low-Earth orbit by transitioning from the International Space Station to other platforms.

Partners

Blue Origin and Sierra Space have partnered with several companies and institutions to realize the project:

Blue Origin: Partner, providing vehicle utility core systems, large-diameter modules, and the reusable heavy-lift New Glenn launch system.

Amazon: logistics and supply chain management. Amazon Web Services: AWS will provide a variety of integrated cloud services and tools to support both near-term and long-term technical requirements including space station development and design, flight operations, data management, enterprise architecture, integrated networking, logistics, and communications capabilities. Sierra Space: Partner, providing Large Integrated Flexible Environment (LIFE) modules, node modules, and runway-landing Dream Chaser spaceplane for crew and cargo transportation.

- Mitsubishi Heavy Industries.
- Boeing: Providing science modules, space station operations and maintenance, and the Starliner crew spacecraft.
- Redwire Space: Providing payload operations and deployable structures, and support for microgravity research, development, and manufacturing.
- Genesis Engineering Solutions: Providing the Single Person Spacecraft for routine external operations and tourist excursions.
- Arizona State University: Providing research advisory services and public outreach through a global consortium of fourteen leading universities.

Starlab

Starlab is the name given to the planned LEO space station designed by Nanoracks for commercial space activities uses. The company released preliminary plans in October 2021. The main structure of Starlab consists of a large inflatable habitat to be built by Lockheed Martin and a metallic docking node. The station is being designed to support 4 persons in 340 m3 of volume. The station also features a 60 kW power and propulsion element, a large robotic arm for servicing cargo and external payloads. The station is supposed to be operational in 2027. The company has partnered with other companies to realize the project:

Nanoracks: Nanoracks owns and operates Starlab and GWC Science Park. Nanoracks, a leader in commercial space services, will respond quickly to customer needs and prioritize on-orbit activities;

Voyager Space: Voyager, the majority stakeholder in Nanoracks;

Lockheed Martin: Lockheed Martin builds spacecraft systems. The company serves as the technical integrator for Starlab and will develop Starlab's inflatable habitat module. After launch, Lockheed Martin will operate the system under Nanoracks' leadership.

Airbus: Airbus builds spacecraft systems and is a major contractor with ESA. The company will provide "technical design support and expertise" for Starlab.

On 2 December 2021, NASA announced it had selected Nanoracks, as one of three companies, to develop designs of space stations and other commercial destinations in space. Nanoracks was awarded $160 million. These Space Act Agreements are the first phase of two with which NASA aims to maintain an uninterrupted U.S. presence in low-Earth orbit by transitioning from the International Space Station to other platforms.

On 4 January 2023, it was announced Airbus would join the Starlab project. "Working with Airbus we will expand Starlab's ecosystem to serve the European Space Agency (ESA) and its member state space agencies to continue their microgravity research in LEO," Dylan Taylor, chairman and chief executive of Voyager Space, said in the announcement.

Concept for a Huge Chinese Space Station

China Wants to Build a Mega Spaceship That's Nearly a Mile Long

China is investigating how to build ultra-large spacecraft that are up to 0.6 mile (1 kilometer) long. But how feasible is the idea, and what would be the use of such a massive spacecraft?

The project is part of a wider call for research proposals from the National Natural Science Foundation of China, a funding agency managed by the country's Ministry of Science and Technology. A research outline posted on the foundation's website described such enormous spaceships as "major strategic aerospace equipment for the future use of space resources, exploration of the mysteries of the universe, and long-term living in orbit."

The foundation wants scientists to conduct research into new, lightweight design methods that could limit the amount of construction material that has to be lofted into orbit, and new techniques for safely assembling such massive structures in space. If funded, the feasibility study would run for five years and have a budget of 15 million yuan ($2.3 million).

The project might sound like science fiction, but former NASA chief technologist Mason Peck said the idea isn't entirely off the wall, and the challenge is more a question of engineering than fundamental science.

"I think it's entirely feasible," Peck, now a professor of aerospace engineering at Cornell University, told Live Science. "I would describe the problems here not as insurmountable impediments, but rather problems of scale."

By far the biggest challenge would be the price tag, noted Peck, due to the huge cost of launching objects and materials into space. The International Space Station (ISS), which is only 361 feet (110 meters) wide at its widest point according to NASA, cost roughly $100 billion to build, Peck said, so constructing something 10 times larger would strain even the most generous national space budget.

Much depends on what kind of structure the Chinese plan to build, though. The ISS is packed with equipment and is designed to accommodate humans, which significantly increases its mass. "If we're talking about something that is simply long and not also heavy then it's a different story," Peck said.

Building techniques could also reduce the cost of getting a behemoth spaceship into space. The conventional approach would be to build components on Earth and then assemble them like Legos in orbit, said Peck, but 3D-printing technology could potentially turn compact raw materials into structural components of much larger dimensions in space.

An even more attractive option would be to source raw materials from the moon, which has low gravity compared with Earth, meaning that launching materials from its surface into space would be much easier, according to Peck. Still, that first requires launch infrastructure on the moon and is therefore not an option in the short term.

BIG SPACESHIP, BIG PROBLEMS

A structure of such massive proportions will also face unique problems. Whenever a spacecraft is subjected to forces, whether from maneuvering in orbit or docking with another vehicle, the motion imparts energy to the spaceship's structure that causes it to vibrate and bend, Peck explained. With such a large structure, these vibrations will take a long time to subside so it's likely the spacecraft will require shock absorbers or active control to counteract those vibrations, he said.

Designers will also have to make careful trade-offs when deciding what altitude the spacecraft should orbit at, Peck said. At lower altitudes, drag from the outer atmosphere slows vehicles down, requiring them to constantly boost themselves back into a stable orbit. This is already an issue for the ISS, Peck noted, but for a much larger structure, which has more drag acting on it and would require more fuel to boost back into place, it would be a major concern.

On the flip side, launching to higher altitudes is much more expensive, and radiation levels increase quickly the further from Earth's atmosphere an object gets, which will be a problem if the spacecraft houses humans.

But while building such a structure might be technically possible, it's not feasible in any practical sense, said Michael Lembeck, a professor of aerospace engineering at the University of Illinois at Urbana-Champaign who has worked on both government and commercial space programs.

"It's kind of like us talking about building the Starship Enterprise," he told Live Science. "It's fantastical, not feasible, and fun to think about, but not very realistic for our level of technology," given the cost, he said.

Given the research project's tiny budget, it is likely only meant to be a small, academic study to map out the very earliest contours of such a project and identify technological gaps, Lembeck said. For comparison, the budget to build a capsule to take astronauts to the ISS was $3 billion. "So the level of effort here is extremely small compared to the outcomes that are desired," he added.

There are also questions about what such a big spacecraft would be used for. Lembeck said possibilities include space manufacturing facilities that take advantage of microgravity and abundant solar power to build high-value products like semiconductors and optical equipment, or long-term habitats for off-world living. But both would entail enormous maintenance costs.

"The space station is a $3 billion a year enterprise," Lembeck added. "Multiply that for larger facilities and it quickly becomes a rather large, expensive enterprise to pull off."

China has also expressed interest in building enormous solar power arrays in orbit and beaming the power back to Earth via microwave beams, but Peck said the economics of such a project just don't stack up. Peck has done some back-of-the-envelope calculations and estimates it would cost around $1,000 per watt, compared with just $2 per watt for energy generated from solar panels on Earth.

10.0 5-15-Years Bases on the Moon

There might be some question about whether a colony on the Moon is living in space but look—it's off the Earth isn't it?

Having moon bases has long been a staple of science fiction. One of my favorite books is called "The Moon is a Harsh Mistress" By Robert Heinlein. It is all about a settled lunar civilization of thousands of people fighting for its independence from Earth.

In recent years Moon circling satellites have detected ice water in the Polar Regions and these discoveries are providing a new impetus for starting a moon base.

Here is more information about Polar bases on the moon:

There are two reasons why the North Pole and south pole of the Moon might be attractive locations for a human colony. First, there is evidence for the presence of water in some continuously shaded areas near the poles. Second, the Moon's axis of rotation is sufficiently close to being perpendicular to the ecliptic plane that the radius of the Moon's polar circles is less than 50 km.

Power collection stations could therefore be plausibly located so that at least one is exposed to sunlight at all times, thus making it possible to power polar colonies almost exclusively with solar energy. Solar power would be unavailable only during a lunar eclipse, but these events are relatively brief and absolutely predictable.

Any such colony would therefore require a reserve energy supply that could temporarily sustain a colony during lunar eclipses or in the event of any incident or malfunction affecting solar power collection. Hydrogen fuel cells would be ideal for this purpose, since the hydrogen needed could be sourced locally using the Moon's polar water and surplus solar power. Moreover, due to the Moon's uneven surface some sites have nearly continuous sunlight. For example, Malapert Mountain, located near the Shackleton crater at the lunar South Pole, offers several advantages as a site:
It is exposed to the Sun most of the time (see Peak of Eternal Light); two closely spaced arrays of solar panels would receive nearly continuous power.

Its proximity to Shackleton Crater (116 km, or 69.8 mi) means that it could provide power and communications to the crater. This crater is potentially valuable for astronomical observation. An infrared instrument would benefit from the very low temperatures. A radio telescope would benefit from being shielded from Earth's broad spectrum radio interference.

The nearby Shoemaker and other craters are in constant deep shadow, and might contain valuable concentrations of hydrogen and other volatiles.

At around 5,000 meters (16,000 feet) elevation, it offers line of sight communications over a large area of the Moon, as well as to Earth. The South Pole-Aitken basin is located at the lunar South Pole. This is the second largest known impact basin in the Solar System, as well as the oldest and biggest impact feature on the Moon, and should provide geologists access to deeper layers of the Moon's crust.

NASA chose to use a south-polar site for the lunar outpost reference design in the Exploration Systems Architecture Study chapter on Lunar Architecture.

At the North Pole, the rim of Peary Crater has been proposed as a favorable location for a base Examination of images from the Clementine mission appear to show that parts of the crater rim are permanently illuminated by sunlight (except during Lunar eclipses). As a result, the temperature conditions are expected to remain very stable at this location, averaging −50 °C (−58 °F). This is comparable to winter conditions in Earth's Poles of Cold in Siberia and Antarctica. The interior of Peary Crater may also harbor hydrogen deposits.

A 1994 bistatic radar experiment performed during the Clementine mission suggested the presence of water ice around the South Pole. The Lunar Prospector spacecraft reported enhanced hydrogen abundances at the South Pole and even more at the North Pole, in 2008. On the other hand, results reported using the Arecibo radio telescope have been interpreted by some to indicate that the anomalous Clementine radar signatures are not indicative of ice, but surface roughness. This interpretation, however, is not universally agreed upon.

A potential limitation of the Polar Regions is that the inflow of solar wind can create an electrical charge on the leeward side of crater rims. The resulting voltage difference can affect electrical equipment, change surface chemistry, erode surfaces and levitate lunar dust.

See my book titled "All About Moon Bases" for many more details on the Deep Space Gateway, and the history and plans for new Moon bases.

11.0 5-15 Years-Hotels in Orbit

Here is an example of an inflatable space hotel made from docked inflated modules:

An article on space hotels:

From Market Research we know that most people would like to stay in orbit for a few days or more. And this stands to reason, if you're paying $20,000 for your trip to orbit! So in order for space tourism to reach its full potential there's going to be a need for orbital accommodation - or space hotels. These will grow through phases, starting with 'lodges' for up to about 100 guests, growing to true hotels of several hundred guests, and eventually to orbiting "theme parks" for many thousands of guests.

Getting There is Half the Fun

But what would a space hotel actually be like to visit? Hotels in orbit will offer the services you expect from a hotel - private rooms, meals, bars. But they'll also offer two unique experiences: stupendous views - of Earth and space - and the endless entertainment of living in zero-G - including sports and other activities that make use of this. And there are further possibilities such as space-walking.

So a trip to a hotel will start with launch to orbit, which takes about 5 minutes of powered acceleration, followed by up to a few hours of weightlessness approaching the hotel (depending on the flight schedule). Docking will be rather like an airliner parking at an airport - but you'll leave the cabin floating in zero-G along the access tube, holding on to a cable with your hands!

The hotels themselves will vary greatly - from being quite Spartan in the early days, to huge luxury structures at a later date. It's actually surprising that as late as 1997 very few designs for space hotels have ever been published. (cf Shimizu, Ehricke, WATG) This is mainly because those who might be expected to design them haven't expected launch costs to come down far enough to make them possible.

Shimizu Corporation's popular Space Hotel design

Luckily it's easy to design basic accommodation in orbit - because it was already done in 1973(!) with the "Skylab" space station. Minimal living facilities require a cylindrical module with air-conditioning, some windows, and a kitchen and bathroom. But zero gravity allows you to build almost any shape and size, in almost any direction. So exploiting the full range of possibilities of zero gravity architecture will keep designers happy for decades! There'll also be rotating (and tethered) structures giving artificial gravity.

Getting Around

Lots of people who've been to space have described in detail what it's like to live in zero gravity. Of course, no-one has yet lived in a rotating space station like the "2001" space station. Such designs will probably be used, but building such a rotating structure will be a significant step beyond just attaching some modules together. It has the advantage of providing accommodation at different levels of artificial gravity, but with some important caveats as discussed by Dr. Theodore Hall.

The key to moving in zero-G is to think of your center of mass - which is just behind your belly-button. Any time you push against someone or bump something, if the line of that push doesn't go through your belly-button, then it tends to set you rotating around your belly button! So to move in a certain direction you have to be sure to push in a line that goes through your belly-button (if you see what that means!)

At first, the key is just to move slowly and simply, so you have time to think what you're doing. But as you get the hang of it you'll find it enjoyable to push off from a wall with just the right rotation rate to land on your feet on the opposite wall. And then faster and faster! There are obviously all sorts of possibilities for dancing, gymnastics and zero G sports!

Luckily you don't need to sleep much living in zero gravity, so you'll have plenty of time for relaxing by hanging out (literally!) in a bar with a panoramic window looking down at the turning Earth below, or sitting in a darkened astronomical viewing room listening to a guide explain the sights you can see through the binoculars available, or discoing in zero G, or...

All Good Things...

Of course all good things have to come to an end, unfortunately! And so after a few days you'll find yourself heading back through the docking point to the returning vehicle - though you'll be much more expert at maneuvering in zero G than you were when you arrived! You'll be thinking how soon you can save up enough to get back up again - or maybe you should change jobs to get to work in an orbiting hotel!

12.0 15-40 Years-Mars Colonies

Yes—I know this is a book mainly about Space Habitats, but Mars is also one of the great current objectives of Space exploration and colony planning. The technologies developed for living on Mars will also be useful for Habitats in Space. In this chapter we will discuss several options for settling Mars.
Mars One is an organization that has proposed to land the first humans on Mars and establish a permanent human colony there by 2032.

Mars One consists of two entities: the not-for-profit Mars One Foundation, and the for-profit company Mars One Ventures. The Mars One Foundation, based in the Netherlands, implements and manages the mission. Mars One Ventures holds all monetization rights, including broadcasting rights. The private spaceflight project is led by Dutch entrepreneur Bas Lansdorp, who announced the Mars One project in May 2012.

Mars One's original concept included launching a robotic lander and orbiter as early as 2020 to be followed by a human crew of four in 2024 and one in 2026. Organizers plan for the crew to be selected from applicants to become the first permanent residents of Mars with no plan of returning to Earth.

Partial funding options include a proposed television documentary program documenting the journey. The project's schedule, technical and financial feasibility, and ethics, have been criticized by scientists, engineers and those in the aerospace industry.

In February 2015, the primary contractors on the initial pre-Phase A contracts had completed all studies paid for by Mars One at that time.

Here is the schedule for flights to get ready for manned habitation:

Initial plan	Current Plan	Milestone	Latest status
2015	2018	Candidate pool reduced to 40 astronauts, replica of the settlement built for training purposes.	Candidate pool reduced to 100
2016	2024	The first communication satellite (ComSat), and a Mars One Lander to demonstrate certain key technologies, would be launched.	Contract with Lockheed Martin
2018	2026	A rover would launch to help select the location of the settlement. The second ComSat would be launched to L5 to enable near-24/7 communication.	Not yet contracted
2020	2029	A second rover and six notional modified Dragon capsules and another rover would launch with two	Not yet contracted

		living units, two life-support units and two supply units.	
2021	2030	The autonomous rovers will begin settlement assembly and operations. The Environmental Control and Life Support System (ECLSS) is planned to have produced a breathable atmosphere of 0.7 bar pressure, 3000 liters of water, 240 kg of oxygen, which will be stored for later use, in the habitat.	
2022	2031	A concept that a Falcon Heavy would launch with the first group of four colonists.	
2023	2032	The first colonists were to arrive on Mars in a notional modified Dragon capsule.	
2024	2033	Departure of second crew of four colonists.	
2025	2034	Arrival of second crew on Mars.	
2031	2040	The colony projected to reach 20 settlers.	

13.0 Space Infrastructure Development

After building The Deep Space Gateway and inflatable Space Hotels, we will need a lot more infrastructure and technology solutions to start constructing the larger space habitats. In the following chapters we discuss:

1) What new technologies will we need to start building large space habitats and other huge space constructions?
2) Who are the leading companies today who are working on some of these technologies?
3) Where will we get the materials for large scale space construction projects?

Required New Technologies

Key to developing space habitats is having the technology to launch key components into space and the types of machines which will build the habitats. Since anyone in space has to wear a many layered and expensive space suit, plus the need for lots of infrastructure to host astronaut construction crews it's obvious that many automated construction tools and robots will be needed to build any sizable habitat in space.

Below are some of the key aspects of that technology which needs to be developed for larger habitat construction:

Reduced launch costs

Today in general it costs about $10,000 U.S. to launch a pound of cargo into Earth orbit. SpaceX says it costs them $2,500 per pound on their Falcon 9 rocket and costs will continue to go down with re-usability and as larger rockets become available.

SpaceX says that their BFR rocket which will launch a much larger payload than the Falcon 9 can launch a freight of 150 tons for approximately $50 per pound. This will allow for incredible improvements of what is feasible to launch into Earth orbit.

Inflatable structures

Bigelow Aerospace has developed a very creative inflatable space structure design which can be launched into space and then inflated to provide lots of habitable volume.

The BA2100 module was a design of theirs which is currently on hold, but you can see the possibilities:

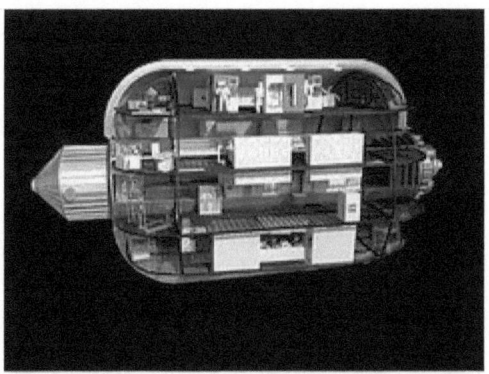

The BA 2100, or Olympus, is a conceptual design for a larger, heavier, and more capable expandable space station module, or interplanetary human transport module, by Bigelow Aerospace. The larger BA 2100 would extend the volume and capabilities of the BA 330 module, which is under development as part of the Bigelow Commercial Space Station. As with the BA 330 module, the number in the name refers to the number of cubic meters of space offered by the module when fully expanded in space.

The weight of the BA 2100 could be as low as 65 to 70 tons (143,000 to 154,000 lb.), but would more likely be "in the range of 100 metric tons".

It is substantially larger than the BA 330, with the docking ends of the module alone estimated at approximately 25 feet (7.6 m) in diameter. The concept model showed the docking ports at both ends. The BA 2100 would require the use of a super-heavy-lift launch vehicle—and would require an 8-meter (26 ft.) fairing for launch, such as the Space Launch System, the Block II version of which would have a 130-tonne (290,000 lb.) payload capacity.

Pressurized volume of single BA 2100 module is 2,250 cubic meters (79,000 cu ft.), compared to 931 cubic meters (32,900 cu ft.) volume of the whole International Space Station as of May 2016.

Currently, a small inflatable module is being tested on the ISS.

Robotic Construction

One of the key technologies which still needs a lot of development work is self-replicating construction robots. Why self-replicating? Because this is the only way to develop a large automated workforce in a reasonable period of time.

The only way to build some very large structures in space would be through a workforce of replicating robots since it would be uneconomical to do the whole job with human construction workers. Human workers would be used more for supervisory work and handling problems which would occur from automated construction activities rather than day to day construction.

The concept of using robots for space construction or an automated spacecraft capable of constructing copies of itself was first proposed in scientific literature in 1974 by Michael A. Arbib, but the concept had appeared earlier in science fiction such as the 1967 novel Berserker by Fred Saberhagen or the 1950 novelette trilogy The Voyage of the Space Beagle by A. E. van Vogt.

The first quantitative engineering analysis of a self-replicating spacecraft was published in 1980 by Robert Freitas, in which the non-replicating Project Daedalus design was modified to include all subsystems necessary for self-replication.

The design's strategy was to use the probe to deliver a "seed" factory with a mass of about 443 tons to a distant site, have the seed factory replicate many copies of itself there to increase its total manufacturing capacity, and then use the resulting automated industrial complex to construct more probes with a single seed factory on board each.

Another NASA study looked at self-replicating robots for space mining:

In 1980, inspired by a 1979 "New Directions Workshop" held at Woods Hole, NASA conducted a joint summer study with ASEE entitled Advanced Automation for Space Missions to produce a detailed proposal for self-replicating factories to develop lunar resources without requiring additional launches or human workers on-site. The study was conducted at Santa Clara University and ran from June 23 to August 29, with the final report published in 1982. The proposed system would have been capable of exponentially increasing productive capacity and the design could be modified to build self-replicating probes to explore the galaxy.

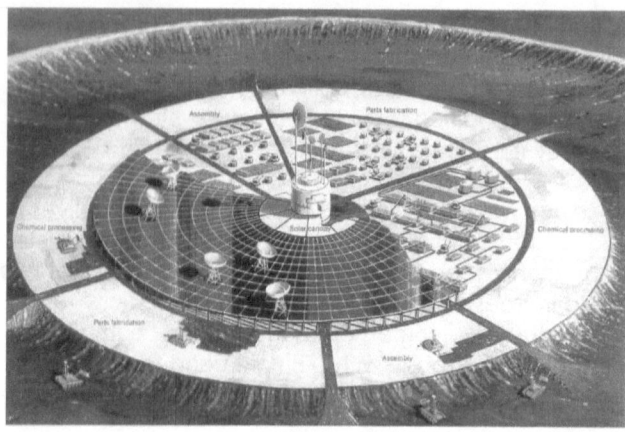

The reference design included small computer-controlled electric carts running on rails inside the factory, mobile "paving machines" that used large parabolic mirrors to focus sunlight on lunar regolith to melt and sinter it into a hard surface suitable for building on, and robotic front-end loaders for strip mining. Raw lunar regolith would be refined by a variety of techniques, primarily hydrofluoric acid leaching. Large transports with a variety of manipulator arms and tools were proposed as the constructors that would put together new factories from parts and assemblies produced by its parent.

Power would be provided by a "canopy" of solar cells supported on pillars. The other machinery would be placed under the canopy.

A "casting robot" would use sculpting tools and templates to make plaster molds. Plaster was selected because the molds are easy to make, can make precise parts with good surface finishes, and the plaster can be easily recycled afterward using an oven to bake the water back out. The robot would then cast most of the parts either from nonconductive molten rock (basalt) or purified metals. A carbon dioxide laser cutting and welding system was also included.

A more speculative, more complex microchip fabricator was specified to produce the computer and electronic systems, but the designers also said that it might prove practical to ship the chips from Earth as if they were "vitamins."

A 2004 study supported by NASA's Institute for Advanced Concepts took this idea further. Some experts are beginning to consider self-replicating machines for asteroid mining.

Much of the design study was concerned with a simple, flexible chemical system for processing the ores, and the differences between the ratio of elements needed by the replicator, and the ratios available in lunar regolith. The element that most limited the growth rate was chlorine, needed to process regolith for aluminum. Chlorine is very rare in lunar regolith.

The current status of this technology in 2017 is an experiment conducted in 2005:

One of the dreams of both science fiction writers and practical robot builders has been realized, at least on a simple level: Cornell University researchers have created a machine that can build copies of itself. Admittedly the machine is just a proof of concept -- it performs no useful function except to self-replicate -- but the basic principle could be extended to create robots that could replicate or at least repair themselves while working in space or in hazardous environments, according to Hod Lipson, Cornell assistant professor of mechanical and aerospace engineering, and computing and information science, in whose lab the robots were built and tested.

Lipson and colleagues report on the work in a brief communication in the May 12 issue of Nature. Their robots are made up of a series of modular cubes -- called "molecubes" -- each containing identical machinery and the complete computer program for replication. The cubes have electromagnets on their faces that allow them to selectively attach to and detach from one another, and a complete robot consists of several cubes linked together. Each cube is divided in half along a long diagonal, which allows a robot composed of many cubes to bend, reconfigure and manipulate other cubes. For example, a tower of cubes can bend itself over at a right angle to pick up another cube.

Although these experimental robots work only in the limited laboratory environment, Lipson suggests that the idea of making self-replicating robots out of self-contained modules could be used to build working robots that could self-repair by replacing defective modules. For example, robots sent to explore Mars could carry a supply of spare modules to use for repairing or rebuilding as needed, allowing for more flexible, versatile and robust missions. Self-replication and repair also could be crucial for robots working in environments where a human with a screwdriver couldn't survive.
To begin replication, the stack of cubes bends over and sets its top cube on the table. Then it bends to one side or another to pick up a new cube and deposit it on top of the first. By repeating the process, one robot made up of a stack of cubes can create another just like itself. Since one robot cannot reach across another robot of the same height, the robot being built assists in completing its own construction.

Each module of the self-replicating robot is a cube about four inches on a side, able to swivel along a diagonal. Self-replicating machines have been the subject of theoretical discussion since the early days of computing and robotics, but only two physical devices that can replicate have been reported. One uses Lego parts assembled in a two-dimensional pattern by moving along tracks; another uses an arrangement of wooden tiles that tumble into a new arrangement when given a shove.

Exactly what qualifies as "self-replication" is open to discussion, Lipson points out. "It is not just a binary property -- of whether something self-replicates or not, but rather a continuum," he explains. The various possibilities are discussed in "A Universal Framework for Analysis of Self-Replication Phenomena," a paper by Lipson and Bryant Adams, a Cornell graduate student in mathematics, published in Proceedings of the European Conference on Artificial Life, ECAL '03, September 2003, Dortmund, Germany.

For example, the researchers point out that human beings reproduce but don't literally self-replicate, since the offspring are not exact copies. And in many cases, the ability to replicate depends on the environment. Rabbits are good replicators in the forest, poor replicators in a desert and abysmal replicators in deep space, they note. "It is not enough to simply say they replicate or even that they replicate well, because these statements only hold in certain contexts," the researchers conclude. The conference paper also discusses the reproduction of viruses and the splitting of light beams into two identical copies. The analysis they supply "allows us to look at an important aspect of biology and quantify it," Lipson explains.

The new robots in Lipson's lab are also very dependent on their environment. They draw power through contacts on the surface of the table and cannot replicate unless the experimenters "feed" them by supplying additional modules.

The research team includes mathematics graduate student Bryant Adams, left, Hod Lipson and mechanical engineering graduate student Victor Zykov.

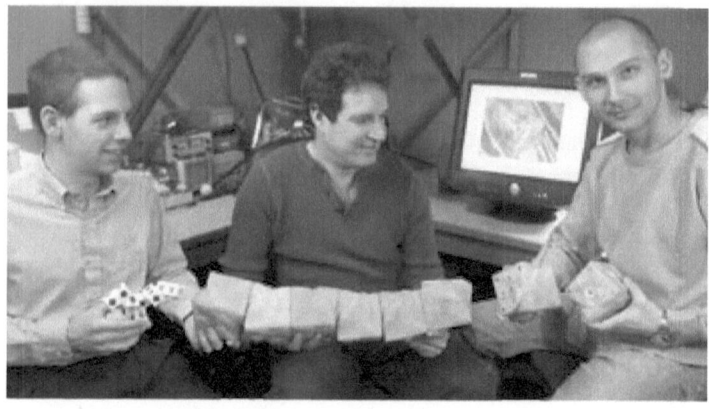

"Although the machines we have created are still simple compared with biological self-reproduction, they demonstrate that mechanical self-reproduction is possible and not unique to biology," the researchers say.

Three Dimensional Printing

Three dimensional printing is a revolutionary manufacturing technology which is overturning accepted manufacturing approaches. In the 21st century it is likely it will become the dominant form of manufacturing. There is already an experimental 3D printer on the ISS.

As 3D printing matures it would be ideal for space based manufacturing of habitats. Again, this is to reduce the amount of human labor in space and to automate many types of materials shaping and forming.

It could be as simple as having 3D printers generate large metal plates for the shell of a habitat, or as complex as embedding electronics inside of materials as part of the overall design.

Solar Smelting

The idea of smelting materials in space is mainly from science fiction since it's never been done. There are some experiments on melting metals in the ISS. Here is a picture of a molten ball of metal magnetically levitated in place:

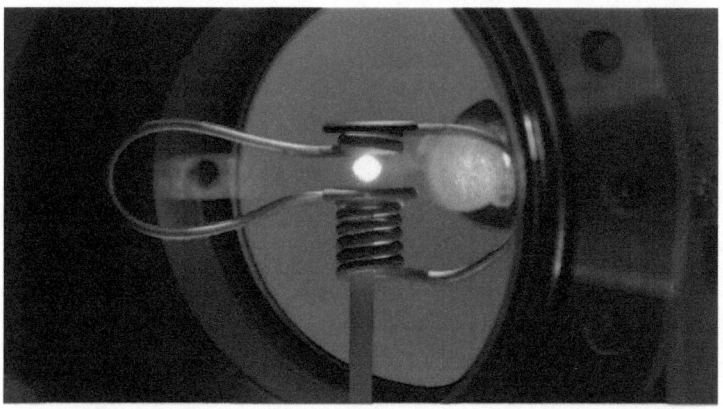

An article on a potential space going solar furnace:

WSMR Solar Furnace could expand spaceflight. TransAstra testing way to drill for water, fuel on asteroids to extend reach of space program

WHITE SANDS MISSILE RANGE - There's something about standing on White Sands Missile Range on a chilly New Mexico morning, watching the power of the sun melt a rock, boring a 2-inch wide by 1-inch deep crater in about one minute, that makes one think of a science fiction scenario.

But that was the reality Friday through a public-private partnership that saw the company TransAstra successfully test technology that, hopefully, will speed human exploration of space by allowing spaceships to drill into passing asteroids to extract both water and fuel.

This would open new horizons in opening our solar system, asteroid belt and eventually deep space by eliminating the need to lift heavy amounts of fluid and fuel out of earth's gravitational field, said Joel Sercel, founder and chief technical officer of TransAstra, who likens to standing between the two towering mirrors that comprise the test facility to the desert, twin-sunned Tatooine of Star Wars fame.

By harnessing the power of the sun in space, large quantities of water could be mined from asteroids and used to fuel spacecraft, thereby reducing the cost of spaceflight as well as increasing distance that can be reasonably flown. In addition to potential economic benefits, Sercel said it is the next necessary step in expanding spatial exploration.

"We are a new company designed to get humans off the planet through asteroid mining," he said.

The challenge is reproducing the results of the testing at WSMR in space. To that end, the researchers have developed a way to test the solar mining technology in equipment that simulates one-1,000,000 of the earth's atmospheric pressure. Next, a project will be designed to send the technology to the International Space Station for further testing, he said.

Propulsion Systems

Currently, chemical rockets are the main type of propulsion available for launching from Earth and traveling the Solar System. Unfortunately, we can't use chemical rockets to go to other planets or visit the Asteroid Belt. It would just be too expensive to shoot all the fuel into orbit and take too long. More efficient propulsion systems are being developed. Here are several:

(See my book titled "Types of Rocket Propulsion and Potenial Space Drives" for more propulsion information.)

Nuclear Propulsion

Nuclear power used for rocket thrust is an approach which has been explored since the 1950s. There are several different approaches. The first is nuclear thermal thrust generated by heating a propellant using a fusion nuclear reactor. The idea is just like how nuclear power plants are constructed where they heat water to steam which runs generators. The main difference would be the reactor would heat a liquid like hydrogen to superheated temperatures and then use the exhaust for propulsive force.

In 2923 NASA and DARPA working together have let a new contract for a nuclear propulsion system. Here is a NASA article about it:

NASA, DARPA Will Test Nuclear Engine for Future Mars Missions

Artist concept of Demonstration for Rocket to Agile Cislunar Operations (DRACO) spacecraft, which will demonstrate a nuclear thermal rocket engine. Nuclear thermal propulsion technology could be used for future NASA crewed missions to Mars.

NASA and the Defense Advanced Research Projects Agency (DARPA) announced Tuesday a collaboration to demonstrate a nuclear thermal rocket engine in space, an enabling capability for NASA crewed missions to Mars.

NASA and DARPA will partner on the Demonstration Rocket for Agile Cislunar Operations, or DRACO, program. The non-reimbursable agreement designed to benefit both agencies, outlines roles, responsibilities, and processes aimed at speeding up development efforts.

"NASA will work with our long-term partner, DARPA, to develop and demonstrate advanced nuclear thermal propulsion technology as soon as 2027. With the help of this new technology, astronauts could journey to and from deep space faster than ever – a major capability to prepare for crewed missions to Mars," said NASA Administrator Bill Nelson. "Congratulations to both NASA and DARPA on this exciting investment, as we ignite the future, together."

Using a nuclear thermal rocket allows for faster transit time, reducing risk for astronauts. Reducing transit time is a key component for human missions to Mars, as longer trips require more supplies and more robust systems. Maturing faster, more efficient transportation technology will help NASA meet its Moon to Mars Objectives.

Other benefits to space travel include increased science payload capacity and higher power for instrumentation and communication. In a nuclear thermal rocket engine, a fission reactor is used to generate extremely high temperatures. The engine transfers the heat produced by the reactor to a liquid propellant, which is expanded and exhausted through a nozzle to propel the spacecraft. Nuclear thermal rockets can be three or more times more efficient than conventional chemical propulsion.

"NASA has a long history of collaborating with DARPA on projects that enable our respective missions, such as in-space servicing," said NASA Deputy Administrator Pam Melroy. "Expanding our partnership to nuclear propulsion will help drive forward NASA's goal to send humans to Mars."

Under the agreement, NASA's Space Technology Mission Directorate (STMD) will lead technical development of the nuclear thermal engine to be integrated with DARPA's experimental spacecraft. DARPA is acting as the contracting authority for the development of the entire stage and the engine, which includes the reactor. DARPA will lead the overall program including rocket systems integration and procurement, approvals, scheduling, and security, cover safety and liability, and ensure overall assembly and integration of the engine with the spacecraft. Over the course of the development, NASA and DARPA will collaborate on assembly of the engine before the in-space demonstration as early as 2027.

"DARPA and NASA have a long history of fruitful collaboration in advancing technologies for our respective goals, from the Saturn V rocket that took humans to the Moon for the first time to robotic servicing and refueling of satellites," said Dr. Stefanie Tompkins, director, DARPA. "The space domain is critical to modern commerce, scientific discovery, and national security. The ability to accomplish leap-ahead advances in space technology through the DRACO nuclear thermal rocket program will be essential for more efficiently and quickly transporting material to the Moon and eventually, people to Mars."

The last nuclear thermal rocket engine tests conducted by the United States occurred more than 50 years ago under NASA's Nuclear Engine for Rocket Vehicle Application and Rover projects.

"With this collaboration, we will leverage our expertise gained from many previous space nuclear power and propulsion projects," said Jim Reuter, associate administrator for STMD. "Recent aerospace materials and engineering advancements are enabling a new era for space nuclear technology, and this flight demonstration will be a major achievement toward establishing a space transportation capability for an Earth-Moon economy."

NASA, the Department of Energy (DOE), and industry are also developing advanced space nuclear technologies for multiple initiatives to harness power for space exploration. Through NASA's Fission Surface Power project, DOE awarded three commercial design efforts to develop nuclear power plant concepts that could be used on the surface of the Moon and, later, Mars.

NASA and DOE are working another commercial design effort to advance higher temperature fission fuels and reactor designs as part of a nuclear thermal propulsion engine. These design efforts are still under development to support a longer-range goal for increased engine performance and will not be used for the DRACO engine.

Nuclear Pulse Propulsion

Project Orion was the first serious attempt to design a nuclear pulse rocket. The design effort was carried out at General Atomics in the late 1950s and early 1960s. The idea of *Orion* was to react small directional nuclear explosives utilizing a variant of the Teller-Ulam two-stage bomb design against a large steel pusher plate attached to the spacecraft with shock absorbers. Efficient directional explosives maximized the momentum transfer, leading to specific impulses in the range of 6,000 seconds, or about thirteen times that of the Space Shuttle Main Engine. With refinements a theoretical maximum of 100,000 seconds (1 MN·s/kg) might be possible. Thrusts were in the millions of tons, allowing spacecraft larger than 8×10^6 tons to be built with 1958 materials.

The reference design was to be constructed of steel using submarine-style construction with a crew of more than 200 and a vehicle takeoff weight of several thousand tons. This low-tech single-stage reference design would reach Mars and back in four weeks from the Earth's surface (compared to 12 months for NASA's current chemically powered reference mission). The same craft could visit Saturn's moons in a seven-month mission (compared to chemically powered missions of about nine years).

A number of engineering problems were found and solved over the course of the project, notably related to crew shielding and pusher-plate lifetime. The system appeared to be entirely workable when the project was shut down in 1965, the main reason being given that the Partial Test Ban Treaty made it illegal (however, before the treaty, the US and Soviet Union had already detonated at least nine nuclear bombs, including thermonuclear bombs, in space, i.e., at altitudes over 100 km: see high altitude nuclear explosions).

There were also ethical issues with launching such a vehicle within the Earth's magnetosphere: calculations using the now disputed linear no-threshold model of radiation damage showed that the fallout from each takeoff would kill between 1 and 10 people. In a threshold model, such extremely low levels of thinly distributed radiation would have no associated ill-effects, while under hormesis models, such tiny doses would be negligibly beneficial. It should be noted that with the possible use of less efficient clean nuclear bombs for achieving orbit and then more efficient higher yield dirty bombs for travel would bring down the amount of fallout caused from an Earth-based launch by a significant factor.

Use of Fusion Reactors

A popular theme in science fiction is to use fusion reactors which have been developed to be light enough to fit inside space ships and power all of the systems.

Fusion at this point is still decades away from commercial use so we don't know how long it would be until it could really be used in spacecraft due to weight considerations even if fusion reactors were already producing power.

Solar Electric Propulsion

NASA's Solar Electric Propulsion (SEP) project is developing critical technologies to extend the length and capabilities of ambitious new science and exploration missions. Alternative propulsion technologies such as SEP may deliver the right mix of cost savings, safety and superior propulsive power to enrich a variety of next-generation journeys to worlds and destinations beyond Earth orbit.

Energized by the electric power from on-board solar arrays, the electrically propelled system will use 10 times less propellant than a comparable, conventional chemical propulsion system, such as those used to power the space shuttles to orbit. Yet that reduced fuel mass will deliver robust power capable of propelling robotic and crewed missions well beyond low-Earth orbit -- sending exploration spacecraft to distant destinations or ferrying cargo to and from points of interest, laying the groundwork for new missions or resupplying those already underway.

NASA's Glenn Research Center in Cleveland leads the Solar Electric Propulsion project for the agency and is preparing a system-level flight demonstration to launch later this decade. Technologies the project is developing and demonstrating include advanced solar arrays, high-voltage power management and distribution, power processing units (PPUs) and high-power Hall thrusters.

During the technology maturation period -- under the auspices of NASA's Game Changing Development Program, prior to transitioning to the Technology Demonstration Missions Program -- the SEP project began developing large, flexible, radiation-resistant solar arrays that can be stowed into small, lightweight, more cost-effective packages for launch. After launch, they unfurl to capture enough solar energy to provide the high levels of electrical power needed to enable high-powered solar electric propulsion.

The SEP project worked with ATK Aerospace and Deployable Space Systems Inc. to build and test two solar array designs: one that folds out like a fan (ATK MegaFlex) and another that rolls out like a window shade (DSS Mega-ROSA). Both use lightweight structures and flexible blanket technology and are durable enough to operate for long periods in Earth orbit or passing through the punishing space environment, including the Van Allen radiation belts.

The SEP project also will use electrostatic Hall thrusters with advanced magnetic shielding instead of a rocket engine with conventional chemical propellant. With SEP, large solar cell arrays convert collected sunlight energy to electrical power. That energy is fed into exceptionally fuel-efficient thrusters that provide gentle but nonstop thrust throughout the mission.

The thruster traps electrons in a magnetic field and uses them to ionize the onboard propellant -- in this case, the inert gas xenon -- into an exhaust plume of plasma that accelerates the spacecraft forward. Several Hall thrusters can be combined to increase the power of an SEP spacecraft. Such a system can be used to accelerate xenon ions to more than 65,000 mph and will provide enough force over a period of time to move cargo and perform orbital transfers.

In fiscal year 2015, researchers successfully tested a new 12.5-kilowatt Hall thruster that employs magnetic shielding, enabling it to operate continuously for years -- a capacity important to deep-space exploration missions. The Solar Electric Propulsion project will demonstrate the key technologies necessary for robotic and human exploration-class SEP transportation systems as well as highly efficient orbit transfer capabilities for commercial space operations and science missions. This effort will benefit not only NASA missions -- such as the Asteroid Redirect Robotic Mission (ARRM), new highly capable science missions and human missions to Mars -- but can provide more affordable primary power and more efficient orbital maneuvering and station-keeping capabilities for commercial communications satellites.

ARRM, the project's planned flight demonstration, will employ a number of Hall thruster strings operating at a total power of 40 kW with a set of solar array wings supplying 50 kW overall.

Later this decade, NASA will demonstrate a Solar Electric Propulsion system in flight, launching a spacecraft to validate the technology and hardware for a high-energy, orbit-transfer mission.

Ion Propulsion

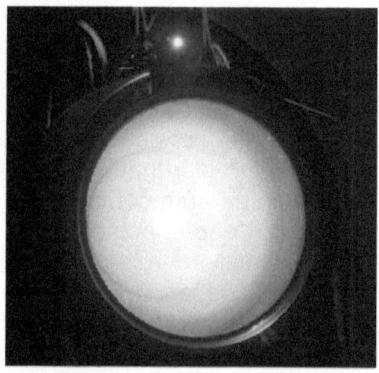

Ion thrusters are being designed for a wide variety of missions—from keeping communications satellites in the proper position (station-keeping) to propelling spacecraft throughout our solar system. These thrusters have high specific impulses—ratio of thrust to the rate of propellant consumption, so they require significantly less propellant for a given mission than would be needed with chemical propulsion. Ion propulsion is even considered to be mission enabling for some cases where sufficient chemical propellant cannot be carried on the spacecraft to accomplish the desired mission.

Current Ion Propulsion-Ion thrusters (based on a NASA design) are now being used to keep over 100 geosynchronous Earth orbit communication satellites in their desired locations, and three NSTAR ion thrusters that utilize Glenn-developed technology are enabling the Dawn spacecraft (launched in 2007) to travel deep into our solar system. Dawn is the first spacecraft to orbit two objects in the asteroid belt between Mars and Jupiter: the protoplanets Vesta and Ceres.

Future Ion Propulsion-As the commercial applications for electric propulsion grow because of its ability to extend the operational life of satellites and to reduce launch and operation costs, NASA is involved in work on two different ion thrusters: the NASA Evolutionary Xenon Thruster (NEXT) and the Annular Engine.

NEXT, a high-power ion propulsion system designed to reduce mission cost and trip time, operates at 3 times the power level of NSTAR and was tested continuously for 51,000 hours (equivalent to almost 6 years of operation) in ground tests without failure, to demonstrate that the thruster could operate for the required duration of a range of missions.

NASA Glenn recently awarded a contract to Aerojet Rocketdyne to fabricate two NEXT flight systems (thrusters and power processors) for use on a future NASA science mission. In addition to flying the NEXT system on NASA science missions, NASA plans to take the NEXT technology to higher power and thrust-to-power so that it can be used for a broad range of commercial, NASA, and defense applications.
NASA Glenn's patented Annular Engine has the potential to exceed the performance capabilities of the NEXT ion propulsion system and other electric propulsion thruster designs.

It uses a new thruster design that yields a total (annular) beam area that is 2 times greater than that of NEXT. Thrusters based on the Annular Engine could achieve very high power and thrust levels, allowing ion thrusters to be used in ways that they have never been used before. The objectives are to reduce system cost, reduce system complexity, and enhance performance (higher thrust-to-power capability).

The EM Drive

The EM Drive is a very controversial space drive. It shouldn't work but according to scientists experimenting with it –it seems to:

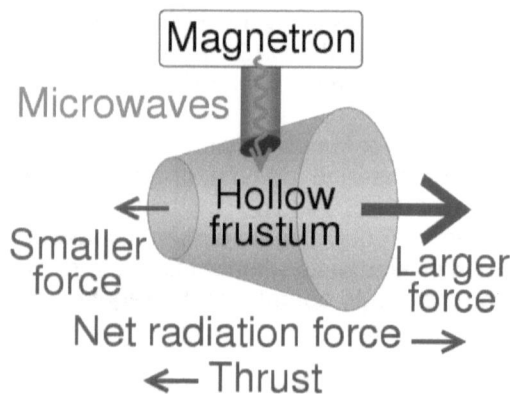

A radio frequency (RF) resonant cavity thruster, also known as an EmDrive, is a controversial proposed type of electromagnetic thruster with a microwave cavity, designed to produce thrust from an electromagnetic field inside the cavity. Skeptics have deemed it to be impossible.

Such thrusters have been hypothesized since 2001 when British engineer Roger Shawyer published details of the concept. To date, there is no theoretical consensus as to how a resonant cavity could produce such thrust. As of 2017, a few tests of prototype drives have observed a small apparent thrust, while other prototypes did not show thrust, and no prototype has been successfully tested more than once.

Due to apparent discrepancies with the laws of physics as they are currently understood and lack of reproducible evidence, many theoretical physicists and commentators have been led to label the device "impossible", explaining the observed thrust by measurement errors. The arising skepticism has led many media platforms to refer to the engine as the Impossible Drive or the Impossible Space Drive

14.0 Leaders in New Technologies

What companies and organizations are the leaders and innovators today in the technologies we will need for building large scale space habitats?

Here are lots of recent major products created by industry to help get to space and build space infrastructure.

(Also see my title "Types of Spaceships: Past, Present, and Future" for more details on new spaceships)

<u>Space X-Rocket History</u>

Space Exploration Technologies Corp., doing business as SpaceX, is an American aerospace manufacturer and space transport services company headquartered in Hawthorne, California. It was founded in 2002 by entrepreneur Elon Musk with the goal of reducing space transportation costs and enabling the colonization of

Mars. SpaceX has since developed the Falcon launch vehicle family and the Dragon spacecraft family, which both currently deliver payloads into Earth orbit.

SpaceX's achievements include the first privately funded liquid-propellant rocket to reach orbit (Falcon 1 in 2008); the first privately funded company to successfully launch, orbit, and recover a spacecraft (Dragon in 2010); the first private company to send a spacecraft to the International Space Station (Dragon in 2012); the first propulsive landing for an orbital rocket (Falcon 9 in 2015); and the first reuse of an orbital rocket (Falcon 9 in 2017). As of March 2017, SpaceX has since flown ten missions to the International Space Station (ISS) under a cargo resupply contract. NASA also awarded SpaceX a further development contract in 2011 to develop and demonstrate a human-rated Dragon, which would be used to transport astronauts to the ISS and return them safely to Earth.

SpaceX announced in 2011 that they were beginning a privately funded reusable launch system technology development program. In December 2015, a first stage was flown back to a landing pad near the launch site, where it successfully accomplished a propulsive vertical landing. This was the first such achievement by a rocket for orbital spaceflight.

In April 2016, with the launch of CRS-8, SpaceX successfully vertically landed a first stage on an ocean drone-ship landing platform. In May 2016, in another first, SpaceX again landed a first stage, but during a significantly more energetic geostationary transfer orbit mission. In March 2017, SpaceX became the first to successfully re-launch and land the first stage of an orbital rocket.

It includes reusable launch vehicles and spacecraft that are intended by SpaceX to replace all of the company's current hardware by the early 2020s, ground infrastructure for rapid launch and relaunch, and zero-gravity propellant transfer technology to be deployed in low Earth orbit. The new vehicles are much larger than the existing SpaceX fleet, and the planned payload of 150 tons (250 tons when flying expendable) makes it a super heavy-lift launch vehicle more powerful than all rockets ever built.

As of the updated of this book in 2023 SpaceX has launched over six manned missions to the ISS and orbit and plans to launch over 100 Falcon 9 Rockets this year.

Spaces has also been launching the Falcon Heavy for several years which uses three Falcon 9 rocket cores to provide the largest heavy lift rocket for commercial launches yet.

SpaceX's New Starship Rocket

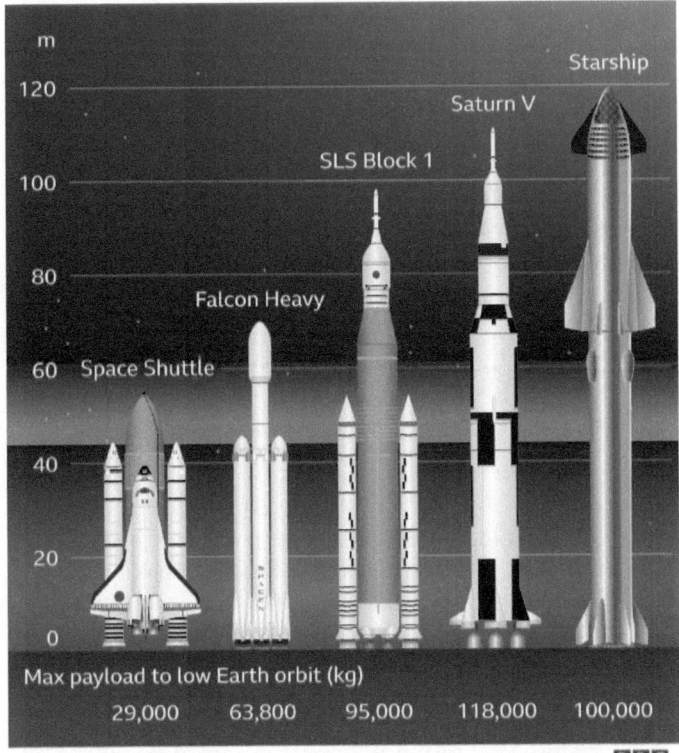

Size Comparison of Major Rockets

The SpaceX Starship is a fully reusable super heavy-lift launch vehicle under development by SpaceX. Standing at 120 m (390 ft) tall, it is designed to be the tallest and most powerful launch vehicle ever built, and the first capable of total reusability.

The Starship launch vehicle is made up of the first-stage booster and the Starship second stage. The second stage functions as a self-contained spacecraft for carrying crew or cargo once in orbit. Both stages are powered by Raptor engines that burn liquid oxygen and liquid methane propellants in a highly efficient full-flow staged combustion power cycle. Both rocket stages are designed to be reused by landing vertically at the launch pad.

In its fully reusable configuration, Starship is planned to have a payload capacity of 150 t (330,000 lb) to low Earth orbit and is designed to be flown multiple times to spread out the cost of the spacecraft. The spacecraft is planned to be refuelable in orbit before traveling to destinations that require more change in velocity (delta-v budget), such as the Moon and Mars. Proposed applications for Starship include regular crewed and cargo launches, building the Starlink internet constellation, and performing suborbital point-to-point flights on Earth.

Development

Starship's development is iterative and incremental, using intensive tests on a series of rocket prototypes. The first prototype, Starhopper, performed several static fires and low-altitude flights. Seven of Starship's upper stage prototypes were flight tested between August 2020 and May 2021. The last of the seven, a full-size Starship SN15, successfully landed after reaching an altitude of 10 kilometers (6.2 mi). A full-scale orbital test flight of the rocket is expected to take place in 2023.

Starship prototype tests can generally be classified into three main types. In proof pressure tests, the vehicle's tanks are pressurized with either gases or liquids to test their strength—sometimes deliberately until they burst. The vehicle then performs mission rehearsals, with or without propellant, to check the vehicle and ground infrastructure. Before a test flight, SpaceX loads the vehicle prototype with propellant and briefly fires its engines in a static fire test. Alternatively, the engines' turbopump spinning can be tested without firing the engines, referred to as a spin prime test.

After successful testing, uncrewed flight tests and launches may commence. During a suborbital launch, Starship prototypes fly to a high altitude and descend, landing either near the launch site, sea, or offshore platforms. During an orbital launch, Starship performs procedures as described in its mission profile. Starship rocket tests, flights, and launches have received significant media coverage.

Blue Origin's Rockets

Blue Origin LLC. is an American privately funded aerospace manufacturer and sub-orbital spaceflight services company headquartered in Kent, Washington. Blue Origin was founded in 2000 by former CEO of Amazon Jeff Bezos, and was led by Rob Meyerson between 2003 to 2017, serving as the company's first president. He was succeeded by Bob Smith who became the company's first CEO. In 2014, Blue Origin moved into orbital spaceflight development as a rocket engine contractor for the BE-4, which suffered delays as a result of a number of technical and managerial problems.

History

Blue Origin was founded in 2000 by Jeff Bezos, the founder and current executive chairman of Amazon. Rob Meyerson joined Blue Origin in 2003 and served as the company's long-time president before leaving the company in late 2018.

The company primarily employs an incremental approach from suborbital to orbital flight, with each developmental step building on its prior work.

Blue Origin moved into the orbital spaceflight technology development business in 2014, initially as a rocket engine supplier for others via a contractual agreement to build a new large rocket engine, the BE-4, for major US launch system operator United Launch Alliance (ULA). Blue Origin said the "BE-4 would be 'ready for flight' by 2017."

On July 20, 2021, the company successfully completed its first crewed mission, Blue Origin NS-16, into space using its New Shepard launch vehicle. The flight was approximately 10 minutes and crossed the Kármán line.

New Shepard performed six crewed flights between July 2021 and August 2022, taking a mix of sponsored celebrities such as Wally Funk and William Shatner, and paying customers. New Shepard ticket sales brought in $50 million through to June 2022. On September 2022, an uncrewed mission of the New Shepard failed due to the failure of the BE-3 main engine. The launch escape system triggered and the capsule landed safely. The remaining New Shepard vehicles were grounded pending an FAA investigation into the incident.

<u>New Shepard</u>

On July 20, 2021, the New Shepard performed its first crewed mission into space. The flight was approximately 10 minutes and crossed the Kármán line. The passengers were Jeff Bezos, his brother Mark Bezos, Wally Funk, and Oliver Daemen, after the unnamed auction winner dropped out due to a scheduling conflict. The second and third crewed missions of New Shepard took place in October and December 2021, respectively. Fourth crewed flight happened in March 2022. On June 4, 2022, New Shepard completed its fifth crewed mission launch after the delayed voyage previous month. The sixth crewed flight took place on August 4, 2022.

New Glenn

New Glenn is a heavy-lift orbital launch vehicle in development by Blue Origin. Named after NASA astronaut John Glenn, design work on the vehicle began in 2012. Illustrations of the vehicle, and the high-level specifications, were initially publicly unveiled in September 2016. New Glenn is a two-stage rocket with a diameter of 7 m (23 ft). Its first stage will be powered by seven BE-4 engines that are also being designed and manufactured by Blue Origin.

Like the New Shepard suborbital launch vehicle that preceded it, the New Glenn's first stage has been designed to be reusable since its inception. In 2021, the company initiated conceptual design work on approaches to potentially make the second stage reusable as well, with the project codenamed Project Jarvis.

Originally aiming for first launch of New Glenn in 2020, Blue Origin has publicly announced delays on three occasions: to 2021 in late 2018, to fourth quarter of 2022 in early 2021, and to no earlier than Q4 2023 in March 2022.

DreamChaser

Dream Chaser is an American reusable lifting-body spaceplane being developed by Sierra Nevada Corporation (SNC) Space Systems. Originally intended as a crewed vehicle, the Dream Chaser Space System is set to be produced after the cargo variant, Dream Chaser Cargo System, is operational. The crewed variant is planned to carry up to seven people and cargo to and from low Earth orbit.

The cargo Dream Chaser is designed to resupply the International Space Station with both pressurized and unpressurized cargo. It is intended to launch vertically on the Vulcan Centaur rocket and autonomously land horizontally on conventional runways. A proposed version to be operated by ESA would launch on an Arianespace vehicle.

Spacecraft

Dream Chaser engineering test article, being driven along the runway after an atmospheric test.

The Dream Chaser design is derived from NASA's HL-20 Personnel Launch System spaceplane concept, which in turn is descended from a series of test vehicles, including the X-20 Dyna-Soar, Northrop M2-F2, Northrop M2-F3, Northrop HL-10, Martin X-24A and X-24B, and Martin X-23 PRIME.

Technology partners

In 2010, the following organizations were named as technology partners for the original passenger Dream Chaser:

Aerojet – reaction control system technology
AdamWorks – composites
Charles Stark Draper Laboratory – guidance, navigation, and control
Lockheed Martin – airframe construction and human rating of the spaceplane
MDA – systems engineering
University of Colorado – human-rating

Propulsion

On-orbit propulsion of the Dream Chaser was originally proposed to be provided by twin hybrid rocket engines capable of repeated starts and throttling. At the time, SNC Space Systems was also developing a similar hybrid rocket for Virgin Galactic's SpaceShipTwo. In May 2014, SNC involvement in the SpaceShipTwo program ended.

After the acquisition of Orbitec LLC in July 2014, Sierra Nevada Corporation announced a major change to the propulsion system. The hybrid rocket engine design was dropped in favor of a cluster of Orbitec's Vortex engines. The new engines would use propane and nitrous oxide as propellants.

Thermal protection system

Its thermal protection system (TPS) is made up of silica-based tiles (for most of the belly and upper portion of the heat shield), and a new composite material called Toughened Unipiece Fibrous Reusable Oxidation Resistant Ceramic (TUFROC) to cover the nose and leading edges.

Crewed version

Artist's conception of the Dream Chaser Space System in the launch configuration of the Atlas V.
The originally planned Dream Chaser Space System is a human-rated version designed to carry from three to seven people and cargo to orbital destinations such as the International Space Station. It was to have a built-in launch escape system and could fly autonomously if needed. Although it could use any suitable launch vehicle, it was planned to be launched on a human-rated Atlas V N12 rocket. The vehicle will be able to return from space by gliding (typically experiencing less than 1.5 g on re-entry) and landing on any airport runway that handles commercial air traffic. Its reaction control system thrusters burned ethanol-based fuel, which is not an explosively volatile material, nor toxic like hydrazine, allowing the Dream Chaser to be handled immediately after landing, unlike the Space Shuttle.

As of 2020, the Sierra Nevada Corporation says it still plans to produce a crewed version of the spacecraft within the next 5 years. The company says it "never stopped working" on the crewed version and fully intends to launch it after the cargo version, and is still committed to the crewed version as of 2021. In November 2021, Sierra Nevada Corporation reported that it received a $1.4 billion investment in Series A funding, which it will use to develop a crewed version of Dream Chaser and fly astronauts by 2025.

On 25 October 2021, Blue Origin and Sierra Nevada Corporation's Sierra Space subsidiary for commercial space activities and space tourism, released their plan for a commercial space station. The station, called Orbital Reef, is intended as a "mixed-use business park". Sierra Nevada Corporation's Dream Chaser was chosen as one of the commercial spacecraft to transport commercial crew to and from the space station, along with Boeing's Starliner.

Artemus Moon Landers

Artist's conception of SpaceX's Starship on the moon

Starship was selected to be NASA's human landing system for Artemis 3 and other landers are under evaluation as additional landers.

Artemis 3 is the third scheduled flight of the Artemis program.

The mission aims to put humans on the moon around 2025, assuming that previous missions of the Artemis program go to plan.

Like the uncrewed Artemis 1 and crewed Artemis 2, this mission aims to launch the gigantic Space Launch System (SLS) megarocket and Orion spacecraft. For landing on the moon, the crew will use SpaceX's Starship, a system that the California company is developing right now for crewed human missions.

The crew for Artemis 3 has not yet been named and the 2025 launch date is rather tentative.

NASA successfully launched Artemis 1 at 01:47 a.m. EST (0647 GMT) on November 16, from Launch Complex 39B at NASA's Kennedy Space Center in Florida.

NASA is awaiting results from both Artemis 1 and Artemis 2 before giving the go-ahead for Artemis 3. Additionally, the agency will require meaningful progress on spacesuits and the human landing system for the moon landing to go ahead.

Related: NASA's Artemis 1 moon mission explained in photos

The Artemis 3 launch date is set for 2025 for now, after an initial target of 2024.

NASA's readiness for Artemis 3 depends on three key things: the success of previous missions, its development of new spacesuits, and the availability of an unflown human landing system or HLS (SpaceX's Starship).

Artemis 1, scheduled to fly in 2022, will assess the readiness of SLS and Orion to carry humans. The major goals of Artemis 2 — scheduled for 2024, if all goes well with the predecessor mission — will then be to fly four humans to the moon's orbit and back. Artemis 2 will last a minimum of eight days (but could extend to as long as three weeks).

Beyond assessing the data collected from Artemis 1 and 2, NASA will also need to make sure its spacesuits are ready. The spacesuits were at first developed in-house by the agency, but it elected to pivot to commercial providers after NASA's Office of the Inspector General warned the agency-made spacesuits were causing undue delay.

The last item will be the readiness of SpaceX's Starship system. Starship has only performed a few tests in mid-air and has yet to fly an orbital mission. SpaceX hopes to fly the orbital mission in 2022. Getting Starship ready for the moon was also delayed due to a complaint (and later lawsuit) by Blue Origin, a competitor in the HLS contract by NASA. The lawsuit is resolved, but it delayed the implementation of the HLS contract by several months.

WHO WILL FLY ON ARTEMIS 3?

The Artemis 3 crew is not yet publicly named. NASA has said the crew will include four astronauts.

NASA astronauts will form at least part of the crew. The agency has pledged the landing crew will include the first woman and the first person of color on the moon (as all of the Apollo human landing missions of the 1960s and 1970s had white men on board.) In 2022, NASA opened eligibility for Artemis missions to the entire astronaut corps.

It is unclear what other agencies may be invited on the trip. Artemis is an international program and a coalition of tens of agencies are forming under the Artemis Accords, which are led by NASA.

The Americans have promised Japan a seat on a future moon mission and possibly, a seat on a landing mission. Japan and the United States came to the pact as part of a larger set of agreements on cybersecurity, 5G cellular networks and other science and technology collaborations, the White House said at the time.

The European Space Agency is also a major collaborator in the Artemis Accords, contributing a European Service Module to the NASA Orion spacecraft as well as several elements to the Gateway space station. A less likely selection would be Canada, as the country already has an astronaut assigned to Artemis 2 in thanks to the country's Canadarm3 robotic technology contributions to Gateway.

WHAT WILL ARTEMIS 3 DO?

Artemis 3 will follow a similar mission profile to Artemis 2, including a launch from the Kennedy Space Center, a translunar injection to bring the astronauts to the moon, an orbiting mission and a return to Earth for a splashdown in the Pacific Ocean.

The mission's major new goal beyond Artemis 2 will be to place two humans at the south pole of the moon, in a zone that is rich with water ice dubbed Artemis Base Camp. (The other two astronauts would remain aboard Orion, in lunar orbit, and the landing area is not yet decided.)

Observations from numerous missions appear to confirm a great deal of water below the moon's surface, particularly in permanently shadowed craters that receive no sunlight. These craters may include hydrogen-rich deposits and water ice underneath the top three feet (one meter) of lunar regolith. Water is a necessity not only for human needs but also for machinery and for growing plants on the surface.

High-resolution maps from NASA's Lunar Reconnaissance Orbiter do provide a great deal of information about the water that is present, although investigators have said there is uncertainty about how pure the water is and exactly where it is located. Additionally, turning water into fuel is still being explored (a CubeSat selected to fly with Artemis 1 is investigating water electrolysis, as a single example.)

NASA is exploring several areas of the South Pole in the coming years using a series of Commercial Lunar Payload Service missions that will place landers, rovers, and scientific payloads upon the surface. Canada also may send a robotic rover to the moon as soon as 2026. The CLPS missions and robotic exploration in the 2020s may assist scientists and engineers with learning how much water is available for humans.

Beyond learning about the moon's water and scientific experiments, NASA expects the landing crew to spend 6.5 days on the lunar surface, which is nearly twice what Apollo astronauts did during their longest missions. They will perform four moonwalks or extravehicular activities lasting about six hours, which is similar to what is performed on the International Space Station. The agency expects the astronauts to use their feet for maneuvering, as they don't have a rover on the manifest.

WHAT COMES AFTER ARTEMIS 3?

The planning past Artemis 3 is highly uncertain given that the mission is relatively far away from us, but NASA does have some early-stage planning for the later 2020s. The timeline for these missions highly depends on how much money the agency receives from Congress, along with the technical progress of the Artemis program.

The agency's "moon to Mars planning manifest" released with its 2023 budget request suggests Artemis 4 will launch in 2027 to help build out Gateway. Then the next human landing on the moon would be Artemis 5 in 2028. The following three landings for Artemis 6 through 8 would happen in 2029, 2030 and 2031.

NASA has been framing Artemis as an opportunity to build out future missions to Mars, but exactly when hasn't been decided yet. In early 2022, the agency said the aim is to focus on the yearly landings on the moon first before targeting the Red Planet in more planning detail.

Planetary Resources-Asteroid Mining

Planetary Resources aims to develop a robotic asteroid mining industry. To achieve this, the company is operating on the basis of a long-term strategic plan.

Unveiling the Planetary Resources 3D-Printed Satellite in February 2014 (Arkyd-300satellite bus configuration). The torus holds the propellant and provides the structure for the satellite.

The first stage will be a survey and analysis, using purpose-built satellites in Earth orbit, to locate the best potential targets among near-Earth asteroids. Several small space telescopes, with various sensing capabilities, are to be launched for this purpose.

The company website asserts that their space telescopes will be made available for hire, for private uses. The company also intends to produce satellites for sale. Their first model of space telescope, the Arkyd-100, has been introduced.

Later stages of the strategic plan envision sending survey probes to selected asteroids in order to map, including deep-scanning, and to conduct sample-and-analysis and/or sample-and-return missions. The company has stated that it could take a decade to finish identifying the best candidates for commercial mining.

Ultimately, their intent is to establish fully automated/robotic asteroid-based mining and processing operations, and the capability to transport the resulting products wherever desired. In addition to the extraction of industrial and precious metals for space-based and terrestrial use, the project envisions producing water for an orbital propellant depot.

3D Printing companies

Three dimensional printing will be needed for large scale space habitat construction to help automate generation of construction materials to reduce human involvement.

There are a variety of companies who are building three dimensional printing machines which can handle different types of metals. Here are the top several three dimensional printing companies in 2017:

1. 3D Systems (NYSE:DDD)
Market cap: $1.36 billion
First on our list of 3D printing companies is none other than 3D Systems. Since 1983, the company has been providing 3D products and services, including 3D printers, print materials, part services and digital design tools. In fact, Chuck Hull, the inventor of stereo lithography, is also the co-founder, executive vice president and chief technology officer of 3D Systems. On that note, the company covers several industries with its products: manufacturing, design and engineering, 3D scanning, and healthcare.

2. ExOne (NASDAQ:XONE)
Market cap: $165.49 million

Founded in 2005, ExOne's business mainly entails of making and selling 3D printing machines and products that are unique to its customers by using installed-base of 3D printing machines. ExOne uses proprietary Binder Jetting technology developed at MIT to print complex parts utilizing industrial-grade materials. This unique heatless technology uses a liquid binding agent that bond layers of material to form an object. Some materials that have been used include metals, sands, and ceramics.

In terms of its systems, ExOne's printers range from production printers, prototyping printers, and research and education printers.

3. HP Inc. (NYSE:HPQ)
Market cap: $36.50 billion
The next 3D printing company on our list is Hewlett-Packard (HP). Although not strictly a 3D printing company, the company is making inroads into the 3D printing market with its Multi Jet Fusion technology. One of HP's focuses is on leveraging 3D printing technology for mainstream manufacturing. In March 2017, the company unveiled the world's first state-of-the-art laboratory in Oregon, as well as a Material Development Kit in collaboration with SIGMADESIGN.

The HP 3D Open Materials and Applications Lab is set to "help companies develop, test and deliver the next generation of materials and applications for 3D printing." Looking ahead, it's reported that the company is looking to expand into the 3D printing metal market, although no further details will be available until 2018.

Relativity Space's New Printed Rocket

Relativity Space made history by launching the first 3D printed rocket into space. Here's how the company aims to beat Elon Musk to Mars by 2024.

A photomontage shows Terran 1 ready to launch on its launchpad next to a picture of the rocket's trajectory after launch.

Terran 1 launched on March 22, 2023, but failed to reach orbit. Trevor Mahlmann/Relativity Space Relativity Space launched the world's first rocket that's almost entirely 3D printed.

The company developed the world's largest metal 3D printers to make its Terran 1 rocket.

It hopes the launch will put it on track to beat Elon Musk's SpaceX to Mars by 2024.

Relativity Space launched the world's 3D printed rocket on March 22, 2023.

Though the rocket failed to reach orbit, it did pass some important milestones on its maiden voyage.

"It was, at least from my perspective, a stunning success, especially with all the variables that they had," Brendan Rosseau, a teaching fellow of space economy from Harvard Business School who wasn't involved in the launch, told Insider.

"They're turning a lot of heads, they're really exciting," he said.

The 3D printed rocket is the brainchild of Relativity CEO Tim Ellis, a former engineer at Jeff Bezos' space startup, Blue Origin.

Here's how Ellis plans to take on his former boss and Elon Musk's SpaceX by disrupting the rocket-manufacturing industry.

Terran 1 is the world's first 3D printed rocket.

Terran 1 is Relativity Space's first functional rocket model. It's also the world's first 3D printed rocket.

85% of the rocket was printed using huge 3D metal printers. The 20,500-pound rocket stands 110 feet high and 7.5 feet wide.

The rocket uses nine custom-built engines to boost it off of the ground and will be able to carry a payload to a low earth orbit of about 2,800 pounds.

The rocket successfully lifted off on March 22, at around 11:25 p.m. ET. Though it didn't reach orbit, it did pass some important milestones to prove the 3D printed structure is viable for flight. You can watch the launch below:

"It's amazing to see how well it performed given that it was the first test launch," Rosseau, from the Harvard Business School, said.

Terran 1 cleared "key points" to prove the 3D printed structure can work, but failed to reach orbit
A picture shows the arc drawn by the rocket's exhaust fumes on the night sky.

The rocket was able to travel far enough that there was no chance of debris falling on the launchpad — passing a technical threshold known as three sigma.

It also survived MaxQ, which is when the pressure on the rocket's structure is at its highest.

Those are two "key moments we're looking to get past on this first flight to definitively prove during flight the printed structures could survive anything we threw at it," Ellis told Insider before the launch.

Still, the rocket, which was not carrying a commercial payload, failed to reach orbit. It successfully separated its first and second stages, but something went wrong shortly after.

"It looks like for the second stage, they couldn't quite get that ignition going, which is always a tricky part," Rosseau said, adding: "That handoff between first and second stage is always challenging when you're trying out a new rocket,"

The upper stage failed to reach orbit, at about three minutes into the flight.

"No one's ever attempted to launch a 3D printed rocket into orbit, and, while we didn't make it all the way today, we gathered enough data to show that flying 3D printed rockets is viable," Relativity test program manager Arwa Tizani Kelly said in a live stream of the launch.

"Obviously I think they would've loved to get to orbit," Rosseau said.

The rockets are built by the world's largest 3D metal printers.

1. Relativity Space

In order to print such a big object, the first step for Relativity Space was to design 3D metal printers that could build a whole sections of a rocket.

The printers, called Stargate, need to print the rocket in about 1,000 pieces. The biggest pieces they're printing are about 20 feet tall, said Ellis.

Here's Relativity's third iteration of the printer putting together a stage of the rocket:

"The largest printers that existed when we started the company could only do about a single cubic foot," said Ellis.

"What Relativity had to do was invent the world's largest metal 3D printers."

The pieces are then joined up by a machine "that's very similar to a 3D printer but it ends up, joining those pieces together again without fixed tooling," said Ellis.

The Terran R, which will dwarf Terran 1, is due to be the biggest-ever 3D printed rocket.
A 3D model shows Terran 1 and Terran R side by side on a black background. TerranR looks about twice as big as Terran1.

A rendering shows what Terran R would look like next to Terran 1. Relativity Space

Though Terran 1 is Relativity's biggest rocket to date, it's relatively small compared to SpaceX's Starship, for instance. Starship aims to carry 150 metric tons into orbit, then return to Earth to be reused.

But for Ellis, the Terran 1 launch is just a step toward his real goal, to build a much bigger rocket called Terran R.

Terran R is also meant to be reusable and should carry 20,000 kilograms of payload. That's about the same as SpaceX's Falcon 9.

It should also be at least 95% 3D printed.

The company is planning to launch this rocket in 2024. This is the rocket Ellis wants to send to Mars.

Though it's still being developed, Ellis is confident it will be put together quickly. AEON R, the engine that will propel the second stage of the rocket forward, has already been tested, he said.

"That was a blank sheet of paper about a year and a half ago," he said.

"So to go from a blank sheet to build the first full of the engine, which we just completed, and then already doing engine component testing at full scale is quite incredibly fast," he said.

Relativity wants to beat Musk's SpaceX to Mars.
An artist's rendering shows Terran R leaving Earth.
An artist's impression of Terran R leaving Earth.

Relativity has teamed up with Impulse Space, founded by former SpaceX propulsion CTO and co-founder Tom Mueller, to set an ambitious timeline for its products: it aims to send TerranR to Mars by 2024.

If it succeeds, Ellis would beat Elon Musk's SpaceX to the red planet, and the mission would be the first-ever commercial mission to Mars.

"2024 is a hyper-aggressive date, there's no question," said Ellis.

"But our pace of execution on TerranR has been quite rapid," he said. "I think that is gonna be key to our ability to execute as fast of a timeline as we can."

Under the agreement with Impulse Space, the payload, a Mars rover, must be delivered by 2029 at the latest, said Ellis.

"I think a big sign that we're committed to this mission and definitely going to make it happen," he said.

Relativity Space wants to be for rockets what Tesla was for electric cars, says Ellis.
Tim Ellis, a brown-haired bespectacled man in his 30s and CEO of Relativity Space, is shown smiling against a grey backdrop, dressed in all black.
Tim Ellis, CEO of Relativity Space. Relativity space
Ellis said his eureka moment came while working at Jeff Bezos' space firm, Blue Origin, as an executive.

Looking around the factory floor, he said, he realized that the company was upheaving the rocket business, but one thing hadn't changed much: the assembly line.

"I saw that we had this giant factory full of fixed tooling, building a whole rocket one at a time by hand with hundreds of thousands to millions of individual piece parts," Ellis told Insider.

"That's really why I started the company to 3D print a whole rocket," he said.

Ellis decided he would disrupt the rocket business by putting metal 3D printers at the core of the manufacturing process.

"The way you have to design for a nearly entirely 3D printed rocket is very different," he said.

"For us, starting from scratch and keeping true to this vision of part-count reduction, designing from the beginning for 3D printing, has led us really leapfrog a lot of other people in this space," he said.

The idea quickly gained a lot of support, of at least $1.3 billion dollars, not least from Mark Cuban, who reportedly responded within five minutes to Ellis' pitch email with $500,000 of the company's seed round.

Ellis says he's revolutionizing the rocket-making industry with his approach.

"It's more similar to what Tesla did with the shift from gas internal combustion engines to electrification, where they realized that you can't just take batteries in electric motors and shove them into a Ford or a Nissan on a traditional manufacturing line," he said.

The printers are flexible, quick, and smart.

The printers can put together a rocket in 60 days, according to Relativity Space's website. Competitors will take between one and two years to build a rocket, per the website.

Because the printers are so quick, it's easier to test different versions of the rocket, which is one reason why Relativity's development has been so quick.

The printer also learns from its mistakes, said Ellis.

"While it's printing, we're collecting many gigabytes and even terabytes of data per print," he said.

"We're also using data science and starting to use machine learning and other more sophisticated data-science techniques in order to have the printers learn from their own prints," he said, adding: "Essentially the printers and the process that the team is going through are getting smarter the more hours we print across our increasingly large fleet of printers."

That also helped the company develop its own aluminum alloy.

"That would only be possible because we're 3D printing. So it's a very integrated process between the design of the rocket and the design of the materials and 3D printers," said Ellis.

The rocket's engine, called AEON, is also 3D printed.
A AEON rocket engine is being fired. on a static test. This engine is entirely 3D printed.
An AEON rocket engine being tested in a static test. Relativity Space
Relativity Space also designed rocket engines that are entirely 3D printed.

These engines are called AEON 1 and AEON R, AEON VAC. They all use a mix of liquid oxygen and liquid methane to propel themselves.

Terran 1 has nine AEON 1 engines powering its second stage. You can see the engines being fired here:

The company has recently launched its Stargate 4, its most advanced printer yet.

Relativity Space's new printer Stargate 4 can print huge components vertically. Relativity Space.

To scale up its production, Relativity Space recently released the latest iteration of its printer, called Stargate 4.

"That is brand new, or at least publicly brand new — it's been in development for a while now," said Ellis.

The difference is that this printer prints horizontally. "That will let us build significantly longer and larger single-piece sections with fewer joints," he said.

It should be able to print objects up to 120 feet long and 24 feet wide. It's also much faster — about 7 to 12 times faster than its predecessors.

The printer is a big part of being able to print Terran R, said Ellis.

Asked if the main goal is to print a rocket in one piece, Ellis said it could be possible, but probably would be impractical.

"Once you get to Terran R scale, you're talking about a vehicle that's well over 200 feet tall. You actually gain iteration speed and build speed by having multiple printers working in parallel," he said.

Ellis sees his race to Mars as more of a collaboration than a competition.

Though Ellis is keen to keep to his competitive deadline to get to Mars, he sees the competition as more of an opportunity for collaboration than a hindrance.

"As far as SpaceX goes I'm absolutely a fan of what they're doing. In fact, the landing rockets and docking at the International Space Station seven years ago when they were a 13-year-old company is one of the things that inspired me to start Relativity," he said.

SpaceX's upcoming planned Starship launch is a big part of NASA's plan to return to the moon, and ultimately Mars.

"It's really important they'll succeed and I do think they will succeed," said Ellis.

Rockets are only one of the products Relativity wants to build. Once it can demonstrate it can bring a payload to Mars, the next step is to bring its printers there.

"It was very clear somebody had to be the second company to join this mission, and that somebody also would need to build an industrial base on Mars. And I think that's gonna be built with 3D printers," he said.

"At first that may be spare parts and other small replacement components for things that break down once you're there. But I think it'll also start to really build out a lot of the food storage, water storage, other industrial equipment that you need initially to sustain kind of the early seeds of people there," he said.

This launch taught the company lessons it'll use to make its bigger rocket.

<u>Relativity Media</u>

Ellis, now in his early thirties, told Insider before the launch he was ready to see his first rocket take flight.

"People are quite pumped, especially given this is such a unique launch with so many firsts, for not just Relativity, but for the industry," he told Insider before the launch.

Among the firsts tested in this launch were the first launch of a 3D printed rocket and the first flight test of methane-oxygen fueled engines.

Though Terran 1's launch is a "huge moment for us," Ellis said, the company will now focus its efforts on building Terran R.

"It's really key that we're proving the 3D printing technology works in flight," he said.

"We will take those lessons learned and then transition those into Terran R," he said.

15.0 Materials for Construction

Where will we get the metals, plastics, and other things we will need for space habitat construction?

Even with much lower launch costs it will still be too expensive to launch simple materials such as steel plates for habitat construction into Earth orbit.

This means that our materials will have to come from other sources which don't have Earth's big gravity well. Likely sources are the Moon and Asteroid belt.

The Moon because gravity there is only one sixth of Earth's so there are many low cost ways materials can be lofted into orbit.
The Asteroid Belt because there is no gravity well and the effort is all about extraction on the Asteroids themselves or moving them into Earth orbit for further processing.

Mining the Moon

Here is a graphic from the Jet Propulsion lab about mining on the moon:

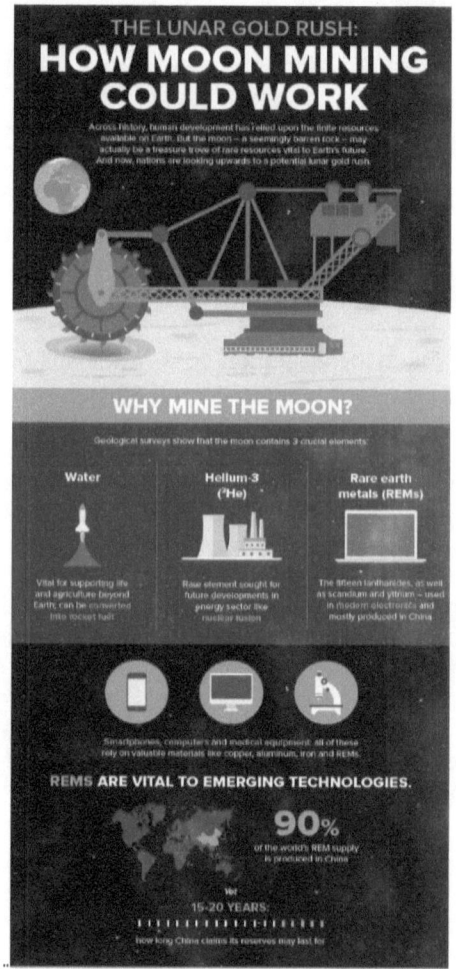

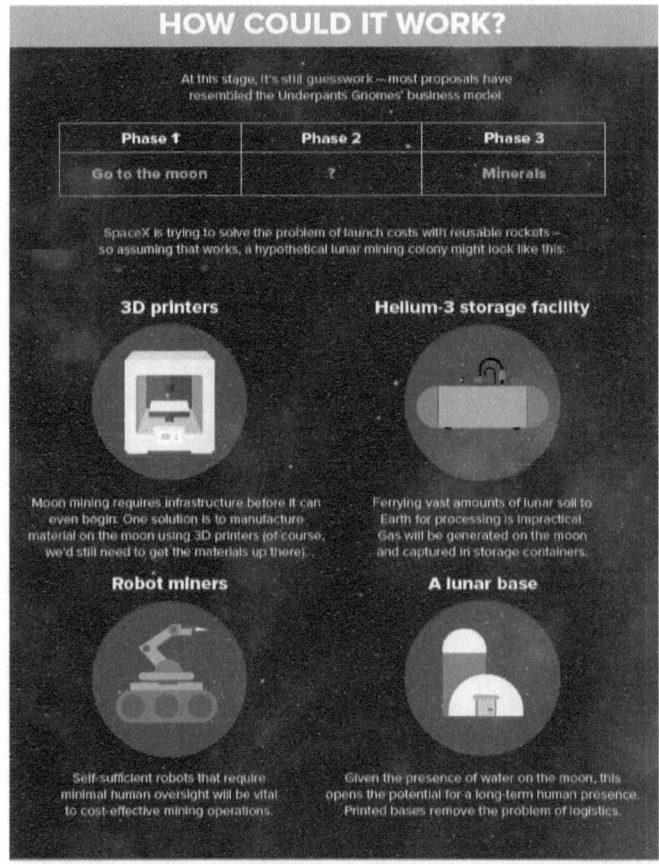

To get materials off the moon several proposals have been made for building mass drivers to take advantage of the moons one sixth Earth gravity:

Consider here a circular mass driver or mass accelerator which would keep power requirements low by spreading the acceleration out over many laps of a circular track. The payload could be about 200 kilograms. If there are passengers or cargo available every 110 minutes for rendezvous with a catcher satellite, it can keep constantly busy.

Suitability for passenger service requires a low radial acceleration, 30 meters per second squared (about 3 g's) will do. This in turn requires a large diameter (about 120 miles). The shape of the device is like a very regular volcanic mountain peak with gently sloping sides and a circular crater on top. The accelerator track would run along the vertical wall of the circular crater.

When the payload and carrier reach orbital velocity (1680 meters per second), the payload is dropped tangentially outward over the top of the wall. A counter weight may be required on the carrier near the base of the wall to balance the carrier. Since the diameter of the track is 120 miles, there is about one and two thirds miles bulge of the curvature of Luna interfering with line of sight communication from one side of the track to the other.

The plane of the circular track makes a 3.2 degree angle with the surface of Luna. (It's like a slice off of the top of Luna one and two thirds miles thick at the

Pole.) Payloads launched tangentially from the track, however, deviate from that plane by curving downward toward Luna in an orbital path. This makes it more likely than otherwise that a payload would smash into a mountain peak. So the accelerator track should be built up on fill as high as practical and care should be taken in choosing the exact direction of launch.

The circular accelerator should be centered at the North Pole while the catcher satellite would orbit about once per 110 minutes at an inclination of about 86.8 degrees. So it would pass over one or another spot on the circular track with every orbit as Luna rotates under the orbit. It could catch a payload whenever a mountain peak did not interfere. Troublesome peaks could be razed.

The above specifications would require 43 kilowatts average power put constantly into payloads plus power to accelerate the carrier and allow for the losses in magnetic levitation. Unfortunately, the carrier cannot constantly accelerate because it must come to a stop to be ready to pick up the next payload. Two tracks, the second with 10 meters less radius and 2 meters more altitude than the first would allow one track to accelerate while the other uses regenerative braking.

As long as we consider developments that must be many years in the future, there is a capability of adding carriers in a train as there is increased available power and need for cargo tonnage. The whole 370 mile circumference of the accelerator could be filled with one train of carriers. The payloads could be connected by rope and the whole train of payloads launched from one point on the circumference into one orbit as 5 minutes and 53 seconds go by.

Asteroid Mining

Mining the Asteroid Belt has been a fixture of science fiction for decades. Now we are getting closer to defining the details of what would be involved in accomplishing this. Here is some information on this mining approach:

Based on known terrestrial reserves, and growing consumption in both developed and developing countries, key elements needed for modern industry and food production could be exhausted on Earth within 50–60 years. These include phosphorus, antimony, zinc, tin, lead, indium, silver, gold and copper. In response, it has been suggested that platinum, cobalt and other valuable elements from asteroids may be mined and sent to Earth for profit, used to build solar-power satellites and space habitats, and water processed from ice to refuel orbiting propellant depots.

Although asteroids and Earth accreted from the same starting materials, Earth's relatively stronger gravity pulled all heavy siderophilic (iron-loving) elements into its core during its molten youth more than four billion years ago. This left the crust depleted of such valuable elements until a rain of asteroid impacts re-infused the depleted crust with metals like gold, cobalt, iron, manganese, molybdenum, nickel, osmium, palladium, platinum, rhenium, rhodium, ruthenium and tungsten (some flow from core to surface does occur, e.g. at the Bushveld Igneous Complex, a famously rich source of platinum-group metals). Today, these metals are mined from Earth's crust, and they are essential for economic and technological progress. Hence, the geologic history of Earth may very well set the stage for a future of asteroid mining.

In 2006, the Keck Observatory announced that the binary Jupiter trojan 617 Patroclus, and possibly large numbers of other Jupiter trojans, are likely extinct comets and consist largely of water ice. Similarly, Jupiter-family comets, and possibly near-Earth asteroids that are extinct comets, might also provide water. The process of in-situ resource utilization—using materials native to space for propellant, thermal management, tankage, radiation shielding, and other high-mass components of space infrastructure—could lead to radical reductions in its cost. Although whether these cost reductions could be achieved, and if achieved would offset the enormous infrastructure investment required, is unknown.

Ice would satisfy one of two necessary conditions to enable "human expansion into the Solar System" (the ultimate goal for human space flight proposed by the 2009 "Augustine Commission" Review of United States Human Space Flight Plans Committee): physical sustainability and economic sustainability.

From the astrobiological perspective, asteroid prospecting could provide scientific data for the search for extraterrestrial intelligence (SETI). Some astrophysicists have suggested that if advanced extraterrestrial civilizations employed asteroid mining long ago, the hallmarks of these activities might be detectable. Why extraterrestrials would have resorted to asteroid mining in near proximity to earth, with its readily available resources, has not been explained.

Some proposed mining techniques for asteroids from the same article include:

Surface mining

On some types of asteroids, material may be scraped off the surface using a scoop or auger, or for larger pieces, an "active grab." There is strong evidence that many asteroids consist of rubble piles, making this approach possible.

Shaft mining

A mine can be dug into the asteroid, and the material extracted through the shaft. This requires precise knowledge to engineer accuracy of astro-location under the surface regolith and a transportation system to carry the desired ore to the processing facility.

Magnetic rakes

Asteroids with a high metal content may be covered in loose grains that can be gathered by means of a magnet.

Heating

For asteroids such as carbonaceous chondrites that contain hydrated minerals, water and other volatiles can be extracted simply by heating. A water extraction test in 2016 by Honeybee Robotics used asteroid regolith simulant developed by Deep Space Industries and the University of Central Florida to match the bulk mineralogy of a particular carbonaceous meteorite. Although the simulant was physically dry (i.e., it contained no water molecules adsorbed in the matrix of the rocky material), heating to about 510 °C released hydroxyl, which came out as substantial amounts of water vapor from the molecular structure of phyllosilicate clays and sulphur compounds. The vapor was

condensed into liquid water filling the collection containers, demonstrating the feasibility of mining water from certain classes of physically dry asteroids.
For volatile materials in extinct comets, heat can be used to melt and vaporize the matrix.

Space Elevators

A Space Elevator would have such a low cost per pound for moving materials to Earth orbit that it would be a viable cost effective way to get habitat building materials into orbit.

The concept of Space Elevators was popularized by Arthur C. Clarke in his science fiction book "The Fountains of Paradise"

The great thing about a space elevator is that it would significantly lower the cost of moving mass to Earth Orbit and initiating spaceflights from the top of the Space Elevator. The problem is that building one will still be a technological leap and project on the order of building an O'Neil space habitat.

The concept of a tower reaching geosynchronous orbit was first published in 1895 by Konstantin Tsiolkovsky. His proposal was for a free-standing tower reaching from the surface of Earth to the height of geostationary orbit. Like all buildings, Tsiolkovsky's structure would be under compression, supporting its weight from below. Since 1959, most ideas for space elevators have focused on purely tensile structures, with the weight of the system held up from above by centrifugal forces. In the tensile concepts, a space tether reaches from a large mass (the counterweight) beyond geostationary orbit to the ground. This structure is held in tension between Earth and the counterweight like an upside-down plumb bob.

To construct a space elevator on Earth the cable material would need to be both stronger and lighter (have greater specific strength) than any known material. Development of new materials which could meet the demanding specific strength requirement is required for designs to progress beyond discussion stage. Carbon nanotubes (CNTs) have been identified as possibly being able to meet the specific strength requirements for an Earth space elevator. Other materials considered have been boron nitride nanotubes, and diamond nanothreads, which were first constructed in 2014.

15.0 50-100 years-Habitat Cylinders

To get a real idea of the construction process for a large space habitat it helps to visualize the whole process.

In my book "Personal Freedom Parts 1 & 2" I have multiple chapters on this building process over a period of five years. Below are those chapters to help you visualize this construction process and how it would look on the inside. (I've left out some chapters which don't deal strictly with the building process or how things look inside the habitat)

Habitat Phase 1-Getting Started

So I ended up balancing two projects at the same time. Project 1 was to start building the habitat. But Project 2 was more difficult.

Project 2 was about building the community who would populate the habitat and how they all became a team

implementing the 10 Principles in their lives in the first place.
Okay—how did we start the habitat?

First it must be realized that we actually had to build two counter rotating habitats linked together so they would not precess out of place. A fancy way of saying that we wanted the two habitats rotations to balance each other out.

The second thing was where to put the habitat?

We decided to use the L5 or Lagrange 5 point which was equidistantly balanced between the earth and the moon. Anything put into this point (or L4) would stay there.

All we would need would be simple thrusters on the Habitats to keep them in place permanently.

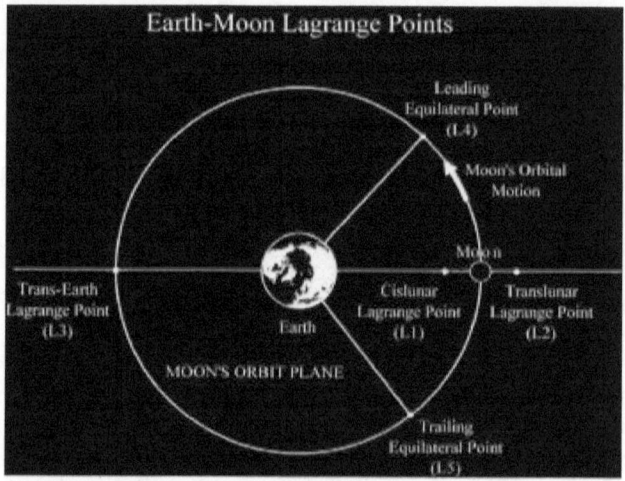

So we assembled a team to build our L5 construction "Shacks" along with tools.

This meant about 20 launches from earth to provide inflatable shelters, supplies, and initial tools.

We also decided we needed about 50 people onsite to get the initial construction done.

Only ten persons were tasked with maintaining our work space station. The rest were involved in checking assembly of our work tools. Let me explain....

Most of the work was to be automated by using large numbers of 3D Printers and space adaptable robots.

Our crew of forty astronauts were really all about helping to assemble additional 3D printers after the parts were created on one printer and about assembling and checking the construction robots.

Some of the work also involved maintenance of both types of tools.

The main habitat(s) were to be five miles in diameter and twenty miles long each. This amounts to about 100 square miles of surface area in each habitat or 200 square miles overall.

Each habitat also had to have three feet thick of titanium and graphene laced concrete to make walls and floors for the structures and to keep people inside safe from cosmic rays or Sun storms.

Speaking of Graphine, where did we get it all? It is a highly manufactured product after all.

Graphine was refined from deposits of graphite sent to us from the Asteroid Belt. The purification work was done at another manufacturing facility from ores they received from the Asteroid Belt.

We bought the purified graphene from one of our partners since it was already being purified in orbit so it was easy to ship it out to us.

The amount of volume to create walls and endcaps for was huge. Way too much you think? We didn't think so…..

Construction was on a geometrically growing scale, because as the 3D printers and robots ramped up-the building process would be going faster and faster.

This also meant the materials we received from the Asteroid belt and our smelting processes had to grow in step.

Our Computers worked out the growth plan-we just had to keep providing assembly and maintenance.

Since each construction plate manufactured was to be 100 foot square—that meant we would need about 600,000 hub plates to be constructed. This really meant 3.6 million plates since we needed three layers of each piece to get an overall thickness of three feet.

Each layer as it came out of 3D manufacturing would be six inches thick so six layers at different angles for the habitat shell.

If there were ten robots welding 100 plates per day it would take over 36,000 days or 99 years. If we had 100 robots 3600 days. We settled on 250 robots or about 2700 days which equals about 7 years.

The build up to 250 robots and about 100 3d Printers would take two years, but that meant the outer hulls of our habitats could be finished in about seven years. So seven years was in the plan to generate the plates, and weld or glue them all together.

The 3D printers we all designed to extrude a surface of up to 100 feet in width and six inches deep. This made for some large panels—but we were in space so the printer just had to be held steady we could make any shape as long as we wanted.

We also had smaller printers which could build electronics and sensors layer by layer. These printers used blueprints to assemble all of the electronics needed for the Habitat with integrated circuits (ICs) being one the few things we didn't produce ourselves.

The 3D printing work was done at the Nano meter scale. The ICs were embedded in the electronics by the printers and the whole electronics were tested then embedded wherever we needed them in the structure.

Panels also had to be solid or transparent, so we used a quartz mix for the see through panels.

I went outside with some of the leaders to look at the robots in action. They were amazing.

Each robot had up to six appendages for grabbing and holding, and were powered by solar cells, with small gas thrusters to allow them to move around and keep on a specific site with other automated workers.

The first thing we had to build though were not panels for the habitat shell, but the structure to hold everything in place.
Imagine an "H" with a much enlarged middle bracket which was over twenty miles in length and the upright of the H was ten miles high. Inside this bracket would fit two "rolling pins" which would be the two linked habitats.

The framework was made of titanium, steel, and graphite to provide the strength needed to hold the habitats in place. The framework also had various station thrusters and docking modules for construction, and future supply ships.

We included tunnels for an eventual electric train network.

Also embedded in the stabilization structure were electronics, thrusters, pipelines, and other things we needed to support the entire two Habitats

We could therefore build each habitat from both ends at once.

The endcaps on each one growing towards each other. You can visualize the end caps starting on each end of the frame work and moving towards each other over time.

While we were getting ready for construction, most of my time as project director was spent making sure our crew was healthy and feeling good about themselves.
We had a lot of social events including celebrations each time we reached a new milestone in the budding construction project.

Also, since the initial phase of construction would take seven years, we had to allow crew members to swap out every six months. Six months at work, and three on the ground, for them to spend time with loved ones, and to fight health and bone density problems from zero G.

Each of us had our own little room-to help us keep our sanity—which was really just the size of an old Apollo Capsule—or the size of several old style phone booths.

The space included a floating sleeping sack, room for personals, and places to stick your sleeping bag on for staying in place. (Chairs aren't needed in zero G.)

Bathrooms were all shared for toilets and showers.
It was while setting up the initial construction effort that I first met Jennifer. She was from the United Kingdom and

was a specialist in programming our Printers and Robots.

She didn't really write the programs since that was done on Earth, but she oversaw how they were running, and made changes to make sure all the 3D printers and Robots were working together in sync.

So we met in one of the weekly Friday evening groups where we all watched a movie or group generated entertainment.

She was about 5 foot 10 inches with a medium build, and blonde with blue eyes. We hit it off immediately and started talking shop about my experiences getting 3D printers to work and hers coordinating many types of units together in space.

We knew each other slightly in a professional manner when she asked me "So Neil what do you do for fun up here?" I hate to admit I was stumped because my whole life was work—I thought work was fun.

I stared at her trying to think of an answer. She saw me gaping and smiled "I see you haven't given the idea much thought. Why don't we think of some fun things to do?" I said "Sure. Would love to."

Thus started our sessions for games twice each week. We mostly played chess but also included other games like Go and some online games too where we could play together.

Sometimes we just went to a windowed port to look at the Earth and Stars—but we agreed not to look at the habitat construction—that was work and off limits.

We were holding each other lightly in the windowed area when Jennifer said "Don't you feel lonely sometimes?" I responded "Yes—But I'm driven to accomplish things—whether with my life or great projects like this Habitat"

I added "But I also like spending time with you. It takes my mind off work, and you make me happy." Jenn liked that and just hugged me tighter.

It was during these observation times when we would hold onto each other which naturally led to some kissing and more… It was nice to have a good friend I could have real "fun" with.

Habitat Building-The Early Years

How can I describe a multi-year project like this? Most of it was drudgery and ongoing maintenance work. Just because we were doing an incredible project in space didn't mean that boredom didn't enter the equation.

Yes—there were accidents like several robots running into each other so Jennifer had to re-program their avoidance algorithms.

Or the time five hundred Habitat outer shell plates had to be replaced due to defects found and to replace them with flawless materials. This really hacked our schedule.

So like everyone else I looked forward to my three months back on Earth.

I hardly ever took a vacation but remembered my Dads old advice: "Either take a planned vacation or you will take the time off anyway" He was referring to getting sick or having serious stress related problems.

Back on Earth I met with my old wellness group back in Houston and we continued to work on our Wellness and our Long Term community health.

One of the keys to our community was that we were all about achieving an enhanced state of mindfulness to all work together in a harmonious manner.

I also coached individuals to work on their personal mission statements and personal goals. This caused many people to start expanding out of their shells. They took up painting, sailing, or even Aikido marital arts.

It was nice to see people coming out of their shells and enjoying life more.

Jennifer and I became closer and she moved into my tiny apartment on Earth so we could enjoy each other's company all the time.

In space we combined two small personal compartments to have a little more space for the two of us.

I really wasn't used to this since I'd always lived alone as an adult.

It also led to that domestic phrase all men fear "Honey, I really need…."

*****

Back at the construction shacks at L5 over the next couple of years, the framework for the habitats was taking shape, and early internal construction was starting on the hubs at each end.

We still weren't up to the full complement of printers and robots but you could see real progress.

Disaster struck when I was off duty. Jenny and I were sleeping in one hanging bag in our combined quarters when the alarms went off and red lights started flashing.

I quickly put on some pants and headed to the control room to see that one of the rocks in a shipment from the asteroids had struck the habitat construction area.

I need to backpedal to explain myself.

The normal process of us getting construction materials was for the Asteroid Mining company to send us balls of rock chunks they had blasted out to a rendezvous point about 100 miles from our site. Then robotic spacecraft would go out to match trajectories and grab hold of the rock blocks or balls to tow them to our solar refinery.

The problem was that the shipment which hit the Habitats was way off course due to a rocket glitch on the rock—which turned out to be a fault in the celestial navigation equipment.

Apparently a contaminated small telescope lenses could give faulty data to the navigation computer.

The rock hit the shell of our under construction Habitat in space instead-while doing several miles per second.

This created a new big hole in the Shelter which endangered the existing construction.

Fortunately the rock was going so fast it just holed the Habitat shell like a giant bullet and left a hole about 200 feet in diameter—which was pretty easy to patch.

The hit had been a type of glancing blow so the other side of the Habitat was not affected.

The hole was easy to fix over the next couple of days, so I was more worried about this happening again rather than the current damage.

The solution turned out to be to install point defense systems on the habitat to send out metal rods to hit and change the trajectories of future objects heading towards the habitat without specific navigation permissions.

The rods were sent by an electromagnetic gun, so they were going thousands of miles per hour when they hit an errant rock or other obstacle.

A time lapse movie would show the end points of the Habitat growing towards each other over years until they finally met.

This was the reason for a huge celebration at our "shacks" and on the ground the Habitat shells finally met together in the middle.

A lot more work remained but this was a major milestone in overall construction.

Redwood Forest-Building the Inside

We focused most of our initial effort on making the eventual Habitat #1 livable as the first one with an enclosed atmosphere. It was where we were going to move our construction headquarters too. We called the first Habitat "Redwood Forest".—(with no actual forest existing at the time)

The name was a statement of our vision—not the current reality.

After we moved our shacks into one end of the Habitat onto the "floor" you could look up to see the rest of the now sterile environment, with very little additional light, and no dirt or constructions on the floor.

We had to use lots of large construction lights around our shacks to keep everything visible.
Our Habitat looked like it had potential, but it certainly didn't look like a home.

The first things we wanted to do were make the "Redwood Forest Habitat" livable.

This meant adding air, heat, and gravity as soon as possible as well as artificial gravity.

GRAVITY

The first issue to solve was how to keep the rotating cylinders in place and what types of bearings would we use to hold the rotating cylinders while they rotated.

We are talking about a structure which has a mass of billions of tons. Something which can be started rotating step by step—but how to keep it in place?

The problem was addressed during construction by using electromagnets to keep the rotating structure separated from the framework by magnetic repulsion.

This also had the benefit of allowing our computers to use sensors throughout the habitat to tell when the weight was not fully balanced, and to put more repulsion into the magnets as a result to keep things centered on the electromagnetic bearings.

Of course one of the main reasons for building the habitats as rotational cylinders was to have artificial gravity which simulated that on Earth.

For a radius of our rotating habitat of 2.5 miles, and a speed of .48 rotations per minute—we would have approximately one gravity under our feet on the floor of the cylinder.
However, we wanted a much lower gravity for additional interior construction.

This meant we set rotation at 0.15 rotations per minute to have a 10% gravity. Enough to keep things in place—but low enough to make moving things around easier.

Since each habitat was built into a spindle with electromagnetic bearings, all we needed were rotational rockets around the outside of the habitat to start speeding up the rotation from no motion to our desired rotational speed.

And the two cylinders had to rotate in opposite directions to keep everything balanced in the whole facility. Computers monitored the opposite rotations and kept everything in sync.

We used solar powered electric plasma rockets to start the rotation. They didn't have to be very powerful and actually took several weeks to get the rotation up to .10 gravities inside from the centrifugal force.

The cylinder would eventually rotate like a top in its holding framework!

AIR

We collected water from ice asteroids in the belt. They were sent to us along with the rock and mineral Asteroids.

We used mirrors to melt the ice into water next to the water holding tanks.

The water also went through filters to remove gravel and other large impurities.

Air was generated from cracking water into Oxygen and Hydrogen. The Hydrogen would be useful as reaction fuel for the station keeping thrusters.

However, Oxygen was also a dangerous flame risk so we imported carbon dioxide from the Earth where it could be compressed and frozen for transport— plus other gaseous elements.

Nitrogen could be collected from cracking some rocks in the Asteroid Belt but the most cost effective collections turned out to be from probes entering the atmospheres of Uranus and Neptune where the probes did brief entries of their atmospheres to collect and compress these gases.

The probes were then sent back to L5 on month's long trajectories to be emptied and then they went back out to collect more gases.

The final composition of the air was similar to the earthly atmosphere at sea level pressures.

Why not go with an air mixture we know works?

HEAT

The eventual plan was to use rotating solar mirrors to send light and heat into the habitat and to simulate a 24 hour day.
This first mirror as also a test. Not only did it take years to build in parallel with the habitat but it had to be polished to a reflective surface by dedicated polishing robots.

The mirror was made of highly polished aluminum we processed from asteroid rocks and was also built by combinations of 3D printers and assembly robots.

We started with one mirror built outside anchored to the Habitat shell, and constantly shining inside to give us maximum heat and light all the time. We could always close windows on our living shacks to go to sleep.

*****

We started pumping manufactured air into the habitat and it took a long time to fill.

In fact for the first few months we had to wear spacesuits inside since the pressure was not much different than the top of Mount Everest as far as our breathing needs were concerned.

Rotational jets were started to get the habitat moving up to .15 rotations per minute. This was done over a few weeks also to watch for problems.

Setting up our temporary quarters was really just about setting up pressurized domes on the shell floor and inflating them. These domes had a lot more room than our previous space based quarters and we each had a decent sized hotel sized room for each of us.

The first morning sleeping in the new dome tent I got Jennifer up and we put on our pressure suits then went outside to see the reflected sunlight on the bare floor of the habitat.

It was an amazing feeling to look at the entire enclosure we were in and it almost felt like being on an Earth which was being built.

We both knew what was needed next were DIRT and WATER.

DIRT

Manufacturing earth like dirt is not easy. It has to have the same consistency as earth dirt, the same minerals, and an organic component of microbes, worms, etc. to be what we called "Live Dirt".

The solution was a mixture of materials from the Asteroid belt which was used by the 3D Printers to turn out microscopically sized modules in the consistency of well-Dirt.

We also needed a process to add living components to the dirt.

We used a mile squared area on the floor of the Habitat to grow earthworms, bugs, and microbes. We grew the "live mix" in smaller batches, then we mixed them in with the large dirt spreads along with some water.

The Dirt all had to be mixed, watered, and living organisms inserted into it.

It usually took several months to become ready and "live", so we had bulldozers moving dirt around all the time to make sure everything was well mixed.

After a few weeks of mixing and "live growth", we used a large balloon and tarp which the dirt was bulldozed onto, then we moved the whole contraption to the desired area and emptied it.

WATER

Back in 2015 it was found there was a huge amount of water in the Asteroid belt with water pooled as frozen ice on Ceres.
It was easy enough to shoot us water ice to melt here at the L5 point.

Water ice was melted then mixed with the dirt and also kept in a large set of bladders in the Habitat.

******

We need a lot of dirt to cover the bare floor of the habitat.
The plan was to provide contour structures for hills and lakes then cover them in dirt an average of twenty feet thick.

Twenty feet was considered a minimum depth to support living dirt and provide enough depth for the roots of trees and plants which would eventually populate the inside.

The plan was to take five years to build supporting structures and add dirt and water as needed.

Then buildings would be built as people started to settle the inside.

*****

We also had to worry about radiation storms from the Sun. Large Solar flares could hurt, people, animals, and our sensitive equipment.

This was partially compensated for by having the Habitat shell be at least three feet think with an average of twenty feet of dirt or water on top.

However, we took one more precaution for large storms—to build real storm cellars.

The Storm Cellars were really built under 30 feet of water, with at least 30 feet above, and 30 feet below. The idea was to build safe rooms with enough Air, Water, and Supplies for a week to house people and animals in the case of the largest solar storms.

These shelters were available to all communities to be used as need. We didn't think they would be really needed except every few years or so.

So visualize that on the shore of many of our large lakes, was a building with stairs going underground.

The building was of course shielded by dirt and plants so it was barely visible—except for emergency signs.

On entering the shelter you would walk down several flights of stairs into the safe room which looked like a hotel with no windows.

Stays were only intended for a few days, so we had dormitory type bunks and apartments for families with communal dining facilities.

The shelters were also designed to be livable with no air outside, and had emergency spacesuits.

These precautions were all in case of a major disaster which might render the Habitat totally airless.

Overall it might take twenty years for the inside our new world to really start to look green and livable, with cities and towns completed so the whole place would feel like communities on earth.

For now though it was our dome structures on the floor and the basic construction and landscaping of filling the empty space which took up our time.

My attention now turned to the community I was building and how to get them ready to settle here.

Growth & Conflict

After three months we rotated back to the Habitat to lead the project on site and see how interior construction was going.

The idea was to start building the interior look and feel from the hub where our construction buildings were and move slowly to the other end.

We were creating many habitable and beautiful living spaces along the way.

The construction might take many years to complete overall.

It was already possible to breathe the air which was now only equivalent to an altitude of 18,000 feet on Earth—still needing supplemental oxygen most of the time, which we carried on our backs, but this was much better than wearing a pressure suit or space suit all the time when "outside" inside the cylinder.

You could already see the forms of hills and depressions for a lake taking shape in the first mile out from our Hub.

Dirt was slowly starting to cover these forms, and water filling in a few depressions. The 10% gravity was good enough to keep everything in place.

Before the dirt was placed in proper shapes-like a golf course- the piping for extensive irrigation was installed. Some areas just had sprays into the air, while other areas needed catch basins underground to collect and move the water to lakes or to be re-pumped elsewhere.

Grass and shrubs were then planted in the dirt since it would all take many years for them to mature.

Some areas were intended to be wild and were planted with an assortment of seeds to mimic the look and diversity from Nature.
We added small animals to the new ecologies too. Larger predators would come later.

*****

Something else was also occurring. Our business partners who we had agreed they could use one quarter of the habitat for what they wanted, for their billions, started building industrial plants on the other end of the facility.

Now, I don't have any problems with industry. We need it to survive as a technological civilization.

What I didn't expect was that these companies would decide to do everything too dangerous to do on Earth here in the Habitat.

Originally they wanted to do research on anti-matter here in the habitat.

When we threatened to reject their funds and possibly close the project down they relented.
An expatriate Russian named Dimitri Mebeken was the CEO in charge for our corporate partners who worked with me as since I was the Habitat project head.

We met on Earth originally and he mentioned some Genetic Engineering work to be done in our Habitat. I didn't know he meant disease causing viruses which needed a Class 4 containment to be safe.

His sponsoring company had already setup their Class 4 Containment Facility at the far end of the Redwood Forest Habitat, and imported the viruses and microbes before we learned exactly what they put in there.

Another "Innocent" sounding manufacturing effort had to do with genetic hybrids of animals which we found out later included introducing human genes into Chimps, Water Buffalo, and Anacondas—among others.

Our source of information was our one watering hole we called the "Alien Alcoholics Anonymous" or AAA where everyone hung out after work to shoot the breeze and drink.

One of the lab workers for the industrial companies spilled the beans. When drunk she related that on the other end of the Hub were technologies over there which would change our Universe.

After plying her with a few more drinks she gave our people all the details.

My security service was small but experienced—it was all written up in a report on my desk the next day.

I called Dimitri to setup a meeting with him the next day and took a small flyer down to his end of the Habitat to meet him.

When I confronted him with the evidence of what was going on he looked me in the eyes and said "We paid billions for this opportunity and we will not have a bunch of delusion fantasy driven idiots ruin what we are doing. The payoff is huge and the risk is minor—everything is safely contained."

"This is what you agreed to for the construction money we provided and Earth has no legal hold on us out here."

"The viruses are in a Class 4 facility and the animals are all penned up in secure areas so they can't get out. "

More famous last words were never spoken

I also had a secondary mission—to get Dimitri interested in the 10 Principles of Personal Longevity and to get him interested if I could.

I gave him a printed copy of the book and told him how it had changed my life.

At first he seemed to be looking for what type of Con I was pulling, but after he saw I was sincere he took the book and said he would read it.

Construction Continues

Our initial Habitat residency became one year since the hull was done, while the inside of Redwood Forest really started to look like someplace.
-
On the outside we now had three fully gimballed mobile mirrors to provide sun everywhere inside in the 24 cycle. It was now as bright at noon here as at midday on Earth.

At our end of the Hub, the construction habitat modules had now moved to inside one of the first true buildings. We did both business and personal apartments in the building.

The building looked well designed and was partly underground so it appeared to grow out of the topsoil. A combination of reflective windows and mirrors help it to blend into the background. A nice trick.

One of the safety features of each permanent structure was that they could maintain atmospheric integrity if the habitat lost atmosphere.

This meant that each building also had a full airlock—although the inner door was left open most of the time.

We considered it extremely unlikely to have a mass depressurization event. Anything likely to make holes big enough to cause us to lose our atmosphere would almost certainly kill the rest of us. But who could argue about having additional safety features in our work places and homes.

I'm all for redundancy and safety. Good thing we did this too or most of us would be dead.

If you looked around at the man made scenery the effect was more and more like being in a munchkin sized wilderness. All the trees were still less than six feet tall although the grasses were looking wild and getting as tall as many trees.

There were now two filled lakes near us. Both more than a mile long and eighty feet at their deepest. They were now being stocked with trout and bass.

We were even building simulated sandy shores so people could swim and sun themselves next to the lakes.

One Sunday Jenny and my friends asked me to try something new for fun. I said I was game. We went to the first Lake shore which was called "Little Mediterranean" and to a shack near the beach.

I was excited to see a small motorboat lifted out of the shack and several pairs of waterskies.

They all knew I used to waterski back on Earth, and I was excited, so I told them I would go ahead and try.

We went to the Lake and I jumped in with the skis and also holding the end of the tow rope which was attached to the back of the boat.

When the boat got up to only 10 miles per hour, I started to rise out of the water. My skis planed on top of the water and I hardly sank in at all. Remember that we were then only at .1 gravities so my body wasn't creating much pressure on the water.

I let one ski go and started doing slalom over the wake of the boat, but everything was too easy— since we were riding high and there was very little resistance from gravity on the water.

The others tried too, so at least we did have a fun day.

The air pressure was now the equivalent of 10,000 feet on earth—and many lakes for resorts existed at those altitudes. So although we were out of breath from the exercise, it was well within our abilities.

*****

We didn't know it then but many experimental viruses and microbes had escaped containment on the other end of the Habitat and a few of them liked to eat metals and adhesive materials.

It turned out that these viruses had migrated through the air to uncovered parts of the structure-especially the transparent parts of the habitat.

Through months they had been eating the adhesives which we used on the Windows to hold them in place because we couldn't weld them.

Gradually the adhesives were being eaten away and small leaks would occur through the cracks in the sides of the windows.

After many months, one of the windows exploded outward from the growing air pressure and its weakening joints. In minutes more windows followed and soon we had a tornado of air trying to get out.

I was sleeping with Jennifer and immediately heard sirens going off and emergency doors closing.

We immediately got our pressure suits on and went to the main control room. Video monitors showed a disaster in process.

We could see instruments recording the drop in pressure to 100 millibars, and less. The lowered pressure was causing the lakes to flash freeze and most of the carefully located plants we freezing and dying.

A tornado of air was outside and was ripping up dirt and plants in all the areas we had tediously constructed and planted.

The problem was not only how to patch the panel which had exploded outward, but how to keep the adhesives from failing on more of them.

Temporary plastic covers were placed over the holes. Now it was time to make longer term repairs.

We decided to address the problem in two ways. First, we sent out the construction robots to replace failing window panels.

They didn't have too much of a problem with escaping air, because most of it was already gone.

At first I didn't know what had caused the disaster, but soon Dimitri called and shame facedly admitted what he thought about the escaped microbes being the problem.

They hadn't told us about the accident a few months previously in which the microbes escaped and he admitted they should have.

I told Dimitri we would have a serious meeting after immediate problem was fixed.

Once the panels had been replaced, we covered each joint in all the windows with antiseptics to kills microbes, and put additional adhesives. The new adhesives were also of a different type which were much less prone to being eaten and destroyed.

We also embedded sensors in every joint to tell if the glue was starting to fail so we could fix it before another explosive decompression.

An additional safety protection were floating balloons which had the electronics to detect a hole and loss of pressure, and a temporary tarp with glue it would place to be stabilized until a repair robot could arrive.

We were lucky that everyone was inside a building and it was the night shift—or we might have lost a lot of people.

The fish in the lakes were lucky since only the tops of the lakes were flash frozen. The deeper parts of the lakes were okay.

However, we did lose young animals in the breeding pens outside of pressurized buildings. That was a real loss of life.

It took months to replace the lost air, but with the improvements to avoid the same problem in the future and with the new monitoring and warning systems we didn't think this problem would happen again.

It took six months to replace all the air and fix the damage. We started over with planting and new animals.

After the first two years inside the structure the first five miles from our hub had been shaped, dirt and water filled in, and planted.

Animals, helpful insects, and birds were released again and they started to multiply.

Fish were added to all the lakes and started to breed so we could catch free fish for dinners.

We kept some lakes very cold so we could grow lobsters and other cold water breeds in them.

*****

However, the problems with the dangerous things which the research companies could end up releasing were still with us.

In a long drawn out set of negotiations with Dimitri Mebeken we agreed on a new set of rules for us to do regular checks on his facilities and joint regulations which would control dangerous biologicals and chemicals in the future.

We also backed up what we were agreeing to with the potential for us to take physical control of the habitat if it came to that.

Dimitri actually agreed—with reluctance –because his people had invested a huge amount in our venture and they lived here too.

The Habitat being destroyed would destroy most of their companies too, so it was in their best interests to have safety regulations in place.

To my surprise, Dimitri and many of the Corporate workers had become believers in the effectiveness of the 10 Principles.

They even came down to our part of the hub to attend classes and feel the changes in themselves from doing meditation or energy practices.

*****

We also had another team which had started to work on the inside of Bora Bora 2—the tropical environment. They were a year or two behind us, but making great progress

Permanent Residency

This chapter is all about residency—not permanent habitation, because we believed that things would continue to grow and change—and who knew what things would look like the Habitats in one hundred years.

In year five after we started building the inside of the structure we decided enough developed environment, safety precautions, and interior buildings were in place to start bringing up non construction workers and their families from earth.

First to arrive were the families of our married workers, and they all had new homes assigned to them which could be separated from each other in our new rolling landscape, or in Condos in the new towns we were building. There were a lot of choices.

We also transplanted our longevity centers to keep helping people implement the 10 Principles as well as introduced a course in our schools to teach all the kids the basics of integrating Spirit, Mind, and Body.

Established religions and other social groups were welcome as long as they agreed to live and let live.

That year over 100,000 people came to live with us it was a real milestone.

We now had not just a continuing construction job, but a community—with all that entails.

Most of the adults needed jobs too, so we set them to work building businesses the community needed.

The new residents started everything from Restaurants to Dry Cleaners, and Grocery Stores.

There were even some pet grooming stores. Yes—people could bring their pets too.

Many distributors and trade organizations were also established since we wanted commercial trade with the other space based communities and Earth too.

I certainly don't want to forget the farmers, since they were the backbone of our lives in many ways.

The long term plan was to make us self-sufficient in food. It was way too expensive to haul food out of a gravity well.

This meant many acres of land to cultivate, many type of crops, and even planting orchards for eventual harvest.

Our farmers were some of the first permanent residents, and we maintained and had many of the new residents working on our hydroponic gardens to grow food there too.

What about meat? Many objected, but we saw that cows and sheep could be raised too for meat, and they were kept in fenced in areas.

Slaughtering was done by a Co-Op which the farmers started, and after the first couple of years of residency, fresh steak, pork chops, and chicken joined the menu in many restaurants and homes.

We had pressure shelters the animals could be moved into too in case of a breach of the habitat and the farmers would get a few minutes warning.

Once entrepreneur even planted a grape vineyard even though it would take several years for the grapes to be pressed and the wine matured enough to be drinkable.

But our biggest need was a government for the both Habitats.

We all knew we wanted to form a representative democracy, and as the communities developed they would elect delegates to our legislature.

The legislature grew as the population grew and it was designed more on the lines of a Parliament, with the majority party electing the Prime Minister who would appoint a working cabinet to manage our Habitats.

We did a few things different however, in line with the 10 Principles to reduce problems. We were trying to avoid large egos and having individuals impose their will over others. Egotistical and selfish rulers have been one of the biggest problems in history. How could we change that?

Since we lived in the age of computers and electronics, we made rules on each law the legislature passed.

The rule was those laws which would cost more than 10% of the yearly budget or affected more than 30% of the population had to be ratified by the general population.

This was done by persistent email messages and apps which kept bugging people to register their decisions until they almost had to do it. This kept general population confirmations at a 90% level—even though it might take a week or more to get a consensus decision.

The eventual population goal for both habitats was one million so we put the total legislature maximum at 500 persons, knowing that the size of the constituent's bases would grow over time. Legislative territories had to do with occupations as well as geographical locations.

For example, all the farmers wanted their own representatives in the legislature. These wishes were granted as well as geographic representatives overlapping the farmer's lands.

We also needed a constitution, and so the first job of the delegates –which lasted a few years was to develop a constitution which needed a two thirds vote to be accepted by everyone

Living in Redwood Forest for Five Years

After five years of building our Habitat community this is how things were:

The manufacturing companies had shrunk and sold off most of their land areas so they now only occupied one eighth of the Redwood Forest Habitat.

They were also building their own manufacturing facilities outside our Habitats and within the L5 point to move all of the dangerous processes.

All of the practices which were a danger to the Habitat or residents would soon be removed to safer areas.

The rest of the Forest Habitat had undergone full initial landscaping. You could stand on the "outside" in the Habitat and see many lakes and hills, with grassed covering them, and trees in the process of building forests.

The tallest trees were now thirty feet high and you could get an idea of what the forests would eventually look like.

Multiple cities had grown and they each consisted of thousands of residents. You could see towers in many of them—since there was plenty of room to grow—which could grow even up to two and a half miles high.

Imagine standing at the End Cap in zero gravity where you enter the Redwood Forest Habitat on the inside.

Near the entrance you would see stairs going down into the ground that led the Habitat transportation system.

There were also a couple of trails which led out of the entrance and down into deeper gravity areas.

The trails started as concrete curving path, in and around trees and grass as you descended. After a mile you started feeling the increase gravity and there were some railings on the trail to help you stabilize and get used to the increased gravity. (Which had been set to one gravity at the highest from increased rotational speeds several years ago.) There were even some areas to lie on the grass and look upward.

When you did that you saw the entire inside of the Habitat and it was glorious.

Imagine three strips of land with lakes interspersed and towns and cities going all the way to the other Hub—until they faded away in the distance. The air being moist and there being haze which turned the other end into blurs.

In between the three strips of living area were the windows where the solar mirrors shined in simulating day and night. There were also gradual gradations of light as the day progressed.
It all had many colors and looked like you were in an airplane looking down on everything.

The Redwoods and other trees were all still less than thirty feet, so the forests were still small, but you could tell what they would become by looking at the overall plan.

We even had a river with rapids where you could test yourself Kayaking if you wanted.

*****

Some new sports had arisen too. Since the gravity near the centerline of each Habitat was zero, it was possible launch yourself from the End Caps and fly along the centerline.

Many people learned how to fly and even held soccer like games in freefall. With floating goalposts. Teams were formed and it became a great televised competition.

In case you got too much out of the centerline and were brought down by gravity, you could open a parachute and float down safely.

This new living space led Sharon-one of our physical trainers-- to start a new really tough competition. She was adapting the idea of the Triathlon to our Habitat. Sharon called it the "Greek Space Marathon". (I don't know why—it just sound good)

The first race was amazing and many of the dual Habitat settlers took breaks from work to stand at the sidelines or watch on monitors.

I watched too since I was really excited and curious about how the competitors would navigate the challenges.
The distances were also scaled down since Sharon added a few segments.

The field of competitors at the starting line were thirty people—men and women, since some of the segments needed skills; not just endurance and speed.

The first segment started in the main habitat area and was a twenty mile bike ride in one gravity.
The men pulled ahead of most of the women, but a few women were faster than the lagging men.

The second segment was using Kayaks to go down the river. A few tipped over and those people were fished out and were out of the race too.

Women caught up at this point until the remaining twenty five competitors were evenly interspersed- men and women.

I could see that Sharon with her yellow shirt was racing and she was in the middle.

The third segment was a three mile swim in one of the lakes. A couple competitors were already exhausted and had to be pulled from the water since they could not go on.

The fourth event, with a now reduced field, were twenty one persons and it was a race to the end cap from the lake.

The race was fifteen miles but it was a very unusual course.

The race started as normal, but the gravity kept going down. 80% at mile three, 50% at mile seven, and gravity kept reducing from there.

In the last five miles the competitors weren't running as much as jumping. In one tenth of a gravity some jumps could go up twenty five feet.

This is where the real skill came in and a lot of women started catching up.

If you jumped too high, it would take too long to land and others would pass you. You had to jump on the right trajectory. Not just use force.

More people came down at bad angles and got sprained ankles or generally banged up and had to quit.

This left fifteen competitors at the end cap for the fifth challenge.

I saw that the group included Sharon, but also some younger teens—boys and girls, who had been raised in the Habitat the last five years and were the closest we had to "Natives". These kids had lots of time to practice and they were used to variable gravities.

The fifth challenge was to fly down the centerline of the Habitat for five miles using a pair of wings you would flap through a mechanical linkage to your arms and legs.

Once you reached the end of the line—which was a banner held into place by tiny jets, you would parachute to the ground.

The sixth and final challenge was the skydive, and the first one to land in the target area on the ground was the winner.

Women started to catch up on the men, again because skills were needed to flap the wings and their lighter mass made a difference in flying down the centerline.
What we didn't expect in this first race was that the teenagers started catching up and passing the adults with much more strength and endurance.

As the group got to the end of their flying challenge, we saw them dropping off the wings and skydiving down to the finish.

Sharon was still in the race, although the real competitors had dropped to ten—since some didn't know how to fly well and were lagging far behind.

So it was Sharon and the teenagers-five boys and girls who dropped fast without opening their chutes. They were hoping to gain speed and open their chutes at the end to win.

I should also mention that each competitor wore a safety harness with steam powered safety thrusters. If they went too fast and the computer monitoring sensors decided they were going to crash, it would slow them way down to a safe landing, but they would also be out of the race.

Several of the kids were way too enthusiastic and their thrusters went off—keeping them safe, but they were fouled out of the race.

It was a close finish, but when it was over a girl named Marra who was fifteen years old had won the race, and Sharon came in second.

Only seven people finished the complete race. The rest had been thrown out in various race segments or injured on the way.

There were several hundred people at the party celebrating this inaugural race, and Marra got the Winners Crown—smiling shyly as many teen girls do.

Even those who got injured were sitting down—with their bandages on-- and having a great time with the fresh barbeque and wine.

It seemed that a new sport had been invented and everyone started making plans to attend or compete in next years "Greek Space Race".

*****

Amazingly, one our biggest cash producing industries was tourism. Visitors would stay at new hotels—we even had some hotel chains from Earth, and they would spend a few weeks hiking around, boating in the lakes, skiing, sailing, or generally relaxing.

But one of the most exciting things the tourists did which really excited me, was to attend classes on the 10 Principles. They wanted to learn how to live to their optimum potential and personal freedom, and take what they learned back with them to change their Earth habits.

I could see that teaching people the 10 Principles and how to use them in their lives could have a very long term impact on the rest of humanity.

These Tourists also brought in a lot of money too which didn't hurt our community at all.

*****

Our people continued to practice mindfulness techniques and integration of the 10 Principles into their lives.

The full impact of our communities learning the 10 Principles in their lives were just becoming known to us.

Once thing was a lot less religious strife. People brought their own religions with them and they had been building churches, synagogues, and mosques.

But, since these people realized the commonalities of all religions, they were polite together and many interfaith events were planned. Everyone had respect for others cultures and religions.

We had our disagreements-as expected, but everyone in our communities were involved in how to reach a consensus on issues of importance.

Schools were working out, and children were learning important skills in them and having fun too.

In fact one of the newest initiatives was to start planning a college—one for each habitat.

One was to be an engineering and technical school concentrating on construction technologies and techniques we had learned in building the Habitats.

A key difference in the Engineering School was what and how we teach.

We taught construction not only with basic courses, but with a virtual construction program where students learned to build habitats of their designs while wearing virtual reality suites.

The programs were constructed using videos and computer simulations we took from the actual construction.

Students could use tools as appropriate and even speed up the clock in simulation to see the results from their efforts.

Many a student died virtually while building in space, but they learned a lot in the process.

The other things students learned was how to apply the 10 Principles to their lives as an engineer and working in the construction environment.

They learned how being more spiritually centered would help them to remain sharp and objective in a dangerous construction environment, and they learned about how to develop consensus during meetings by avoiding emotions and looking to the common good and the best decisions objectively.

The other college in Bora Bora Two was to be about Philosophy and Wellness; combining the practical application of the 10 Principles to people's lives.

That school was designed to produce coaches and teachers who had internalized these principles and wanted to help others.

Side courses taught specifics for implementation including Yoga of different types, Meditation Instruction, Tai Chi Teaching, and more.

Bora Bora Two

Talking about Bora Bora Two is fun because I love that place!

Jennifer and I would often take weekend "vacations" there with our friends to see it
We took the tube trains which now connected distant parts of the Redwood Forest Habitat with each other as well as to Bora Bora Two.

These electric and computer controlled trains could take you to stations anywhere in the Habitats within 30 minutes. Trains came to each station every 15 minutes and could each hold 50 people.

We reached Bora Bora Two (BB2) on the train from our station and headed out into the "islands" to the beach house we usually stayed at.

Inside BB2 it was 70% water, with a simulated ocean near the shore. The Ocean was continuing to be stocked with salt water fish and other sea animals including many types of fish and shellfish.

A reef was even in the process of being built and grown to provide a real reef experience to our children and more diversity of life to our Habitat.

Streams were pumped out at the top of hills to have fresh water and waterfalls everywhere too.

The Ocean was up to 100 feet deep for bottom dwelling fish to live and to have a darker experience for some breeds of fish and whales.

Yes-we had whales—a small breeding colony of grey whales, but they added to our diversity and the food chain.

We also had canals to let the whales move from segment to segment to let them move through the seas and all over BB2.

If you hiked some of the islands—which went up to over 1000 feet high-with cliffs, and looked out over the Ocean, you got a real feel for being on a South Seas Island in the Pacific.

We did avoid raising man eating sharks, but did have some of the gentler breeds as part of the overall food chain.

*****

Several fish farms and fishing businesses developed to supply residents and visitors with fresh seafood for their diets.

You could go to a fish market or a restaurant and find many of the choices an earthly fish market would offer.

So imagine us living on the beach outside of our house there, and grilling lobsters along with fresh swordfish steaks together.

Add to that some fresh corn on the cob which we brought with us from Redwood Forest, and some home grown beer, and you have the makings of a little summer feast.

We sat with our friends after being satiated by dinner and watch the man made twilight descend over the ocean. The number of colors on the ocean were amazing.

It was incredible to think that we had created this new world from nothing and in the dark and cold emptiness of space.
People were meant to live in tune with Nature. Even if it was an artificially created one.

But our sense of adventure was not totally fulfilled. What could we do next for a next adventure?

17.0 75-150 years-Asteroid Homes

Asteroids are another types of potential habitat which have advantages and limitations compared to an entirely artificial space habitat.

I wrote about building this type of habitat in my book "Earth Protector-The Psychic Soldier Series Book 4"

These chapters detail building a living space cavern in the center of the small moon Prometheus inside the rings of Saturn:

<u>My Secret Hideaway</u>

What I wanted to find was a secret hideaway off Earth where I could have some real privacy. By the early 21st century there was almost no place on Earth where you could really get away from it all. Even Antarctica had planes and sensors everywhere, as well as the omnipresent spy satellites from each country overhead.

I still had millions from my investment efforts in the early 20th century and with Wall Street financial managers for the last fifty years. Most of the investment was done in conventional businesses through shell companies which I wholly owned. In fact when I checked the value of my companies, it was amazing to see that my overall wealth was above one billion dollars. But I was so well diversified and had such a network of shell companies that nobody would ever be able to find out who I was.

Through my connections and position I was able to convince some U.S. contractors and designers in Arakesh to design me a small ship. The design would be kind of a space going Patrol Boat but also with a big cargo carrying capacity. There were quarters for up to five persons, but the whole ship could be flown by one person in a pinch.

I cashed in a lot of my company stock to pay the initial fees to a construction yard which was also building larger ships for our space going navy in a classified hanger near Atlanta, Georgia. The construction would take one year but there were ways for me to keep busy in the meantime. While I waited for the finished spaceship I searched for the perfect place for my secret personal base.

*****

I was surveying the solar system for someplace where I could have lots of room, be remote from civilization, and pretty well hidden. But I did want it to be close enough that I could return in couple of days if needed for military operations.

Through both internet research and time hired from large Earth telescopes I found several candidates and narrowed them down to one location.

The decision was to use the moon Prometheus in the inner rings of Saturn. It was situated at the inner F Ring in the Saturn system. The moon was shaped like a potato and was about 70 miles long by around 35 miles in diameter. This moon seemed to interact with the F Ring to keep it in shape. For my purposes it was perfect because it was too close to the rings to be safe to visit and I could core into it and build my own little world. It was also a pretty isolated location but within a two day ship journey to Earth.

Finally my ship was ready and I named it the "Stinger" because it was small, but also carried weapons propulsion out of its size. The ship was 150 feet long and 50 feet in diameter with about half taken by cargo space. The engines and life support took up half the space, and half was left for habitation. It also had a warp field generator but the most important to me was what was loaded into the cargo bay.

In the cargo bay I had loaded some A.I. digging and construction machines along with plants and animals. I had big plans for my personal base. The trip out there took two days by ion drive and then I landed on the small moon. Since the gravity was almost nonexistent I used the ship controls and shot steel darts into the moon to anchor. Then I unloaded the construction equipment and set it up outside wearing my spacesuit.

The digging equipment was pretty sophisticated. It used gas jets for orientation and lasers to dig. All I had to do was program directions and it would do the rest.

Disintegrated rocks would shoot out behind it in gaseous form until they froze into particles in the vacuum of space. If there was too much accumulation then robots directed by the overall A.I. would scavenge up the materials and dump them off of the moon. With its negligible escape velocity there just needed to be one toss and the materials would reach escape velocity.

I had also brought manufacturing three D printers which printed more excavation machines from materials I loaded into them from the belt. These additional excavators kept speeding up the process of extracting rock. Similar machines were printed to remove the rock and to turn rock into dirt after excavation was completed. Power was provided by batteries which were re-charged regularly from the nuclear fusion reactor on my ship.

The machines worked 24 hours per day and with multiple excavators and more being built every day, they moved fast. Over the next months the machines dug into the moon down to a depth of 15 miles. Then I had them dig sideways to start excavating a large cavern where I could build my cavern as a nature park, setup my home, and everything else. The cavern was designed to be two thousand feet in diameter and five miles long. It was protected from radiation and any impacts on the moon by 15 miles of rock. I might even have a larger cavern constructed in the future but this was big enough to start with. It was really kind of a private space.

A construction firm was hired after the project got started and they brought in a hundred people to manage most of the machines and speed up construction. The agreement was that all the workers would have to have their memories suppressed after the job was done. They agreed because the wages were five times normal construction wages. The process of hiring these people had some similarities to get soldier recruits used to what we were doing in space by taking them to the moon. I also had a short training program to teach the construction people how to use our advanced construction machines. Temporary living shacks were soon setup all over the cavern.

I brought a couple of other excavators and within six months the cavern was taking shape. Built another entrance to remove materials while I started adding insulation around the excavated volume to keep heat inside since the entire moon was cooled to a couple hundred degrees below zero. This insulation would also would act as a foundation for ground I wanted to plant with trees and bushes.

We also setup thrusters on the outside of the moon to start it rotating to produce centrifugal force for artificial gravity. The gravity took a year to build up to one sixth Earth gravity, then it would keep going indefinitely without any more thrust.

Finally, it was time to put in the airlocks and add atmosphere. Equipment was mined from the rings for oxygen and other gases to make an atmosphere. Then I programmed the excavation machines to breakup rock inside the moon to make soil for plants. Extra ring water made some lakes. After another couple of months of construction and processing the inside cavern was ready for residence.

A fusion generator I'd transported from Earth provided power and large arc lights for the cavern. Construction robots built a large home for me out of concrete. The house was also designed to withstand decompression in case the cavern was ever evacuated. I also imported real wood for paneling and programmed the robots to build a fireplace.

Lastly, trees and bushes were planted all over the cavern and we released some small animals too. These included squirrels, some spade rabbits, and goats. The ponds were also stocked with fish. After a couple of years it would be like walking in nature to walk trails around the cavern.

Over the next few years the cavern was finished off. I took some of my close friends from the space force to visit. My ship landed at the moon into a pit which had been dug into the regolith which had a roof to protect it and provide atmosphere. When we got out of the ship we took rail car down into the cavern. Reaching the cavern after a 10 minute high speed ride their jaws dropped when they saw the size of the cavern and the natural setting we had built.

We walked from the elevator into the central cavern at the point where gravity was zero. A path led down to the rest of the cavern and we took an electric golf cart to go the several miles to my home. Along the way we passed lakes and recently planted woods which I explained would eventually turn into forests. Some animals were proliferating and you could see goats in the fields eating grass.

The house itself was a rambling ranch type house in concrete with large windows and wooden paneling. There were lots of rooms for comfortable living and many bedrooms. Sewage, water, power, and data were all built into the home with processing done in side caves dug under the home.

Overall, I suppose the cavern could house thousands of people, but it was just for me to start with. However, I really didn't want to be a hermit. So, I went back to Earth and to our bases and hired over a hundred men and women for various jobs at my retreat. Everything from cooking to security. Within ten years of starting construction the place was pretty well finished and I was enjoying it very much. Actually, part of what I had enjoyed was the construction process itself.

We even had a little town a couple of miles from my home with houses and stores setup for my employees. Many of the construction workers had decided to stay and run the stores. We needed some more people incentivized to join us to run the stores. We offered them free medical care and schools for kids among other benefits to encourage them to settle their families too. I even had a separate office building constructed next to my home for specialists to track my investments, a personal military staff to keep in touch with allied space force developments, and my own office and conference room.

Ten years from the start of construction, the moon now had almost three hundred persons living and working there in various roles. I wasn't alone, but guess I never really wanted to be totally alone. At least it was very comfortable with lots of social gatherings and parties which I participated in. The single docking pit on the surface had also now been expanded to hold many ships at once. Additional safety airlocks were constructed as well as shelters near the town in case we somehow lost pressurization.

I held many of the parties in my home's large main hall, and they were for local holidays like construction start day, or main habitation day, as well as lots of Earth holidays. We finally had a naming party where I christened our remote settlement as "Tristin Da Cunha of Saturn" It was a play off one of the remotest islands on Earth and my first name.

The idea was still to keep this outpost pretty secret, so we imposed the same rules the military had. Everyone had their subconscious treated to keep from talking about it, and they all signed confidentiality statements to not release this information to anyone outside of the moon.

It was now the year 2016 and I had my new billionaires residence will all of the luxuries I could ever want—except companionship.

18.0 1000+ years Very Large Structures

There are concepts for even much larger structures which will take a comparably larger space based infrastructure and more advanced technologies to develop them.

One direction is to just upsize an O'Neil cylinder. What if we go a cylinder from a length of twenty miles long and five miles in diameter to one hundred miles long and twenty miles in diameter. This type of construction would take stronger materials and increased sizes of systems, but would not be a fundamentally more difficult structure to build.

However, there are construction concepts which would be on the fringes of our imagination:

18.1 Ringworld Structures

The idea of a Ringworld was made popular in the hit science fiction novel by Larry Niven called "Ringworld". I really loved this book because it was about the adventures of a group of explorers who found the Ringworld by accident and their adventures traveling on the Ringworld. Its huge land area, civilizations they found along the way, super strength materials, and much more.

An Article from Popular Mechanics gets into more of the details of building such a structure:

Sci-fi author Larry Niven conjured up such a megastructure for his award-winning 1970 book *Ringworld*. Niven imagined a ring with a radius of 93 million miles—the sun-Earth distance—with the sun placed at the center. The ring' would reach 600 million miles across and a million miles tall. The vast landscape could comfortably support perhaps trillions of humans (or another similarly ambitious, technologically advanced race).

"The thing is roomy enough: three million times the area of the Earth. It will be some time before anyone complains about the crowding," Niven wrote in a 1974 essay entitled "Bigger Than Worlds."

Niven figured a Ringworld would have a thickness of a few thousand feet, and require raw materials with a mass equal to that of Jupiter. Mountain "walls" a thousand miles high would line each rim, preventing the atmosphere from leaking into space. The inner surface could be sculpted like Earth's surface—full of great (though shallow) oceans, soaring mountains, and prodigious farmland—or whatever its builders desired.

Could a Ringworld ever be made? While the concept does not bend physics past the point of breaking, it would require truly extreme engineering and an utter mastery of the forces of nature. According to Anders Sandberg, a research fellow at Oxford University's Future of Humanity Institute who has studied megastructure concepts, a Ringworld "is an amazingly large structure that's way beyond what we can normally imagine, but it's also deeply problematic."

Establishing Gravity

When imagining the ring, Niven had started with the concept of a Dyson Sphere, an idea explored by physicist Freeman Dyson a decade prior to *Ringworld*'s publication.

In its usual science fiction presentation as a "ping pong ball around a star," Niven said, a solid Dyson Sphere lacks gravity. Rotating the sphere would create gravity via centrifugal force, but only the equatorial regions would reap the benefits. "So," Niven tells PM, "I just used the equator."
A Niven Ring, then, can be thought of as the slice of the habitat-friendly section of a Dyson Sphere. To get Earth-like gravity, the Ringworld would need to spin

at nearly three million miles per hour. Very fast, to be sure. But in a frictionless space environment, it could be doable. The ring could work up to that speed over time and then maintain it with little additional thrusting.

Managing the Sun

Although it would be equidistant from its central star at all points, the Ringworld would not, in fact, be gravitationally stable. Any perturbing force from, say, a meteorite strike or a close encounter with another star could throw the Ringworld out of attractive equilibrium and onto on a cataclysmic collision course. "A Ringworld will tend to drift off whenever it gets a chance," Sandberg says.

Readers of the original Ringworld, including students at the Massachusetts Institute of Technology, wrote letters to Niven about this and other technical issues related to the megastructure. Niven addressed the problem in the 1980 sequel, *The Ringworld Engineers*. Large rockets placed along the Ringworld's edge would have to periodically fire to keep the megastructure properly situated away from its sun.

For residents of the Ringworld, that sun would always be directly overhead at a perpetual high noon. To create a day–night cycle and save plant life from frying, Niven envisioned a set of "shadow squares" around the sun at about Mercury's distance from Earth. The parts of Ringworld between the squares would experience roving daylight, while the eclipsed portions would rest in the shade. The whole length of the Ringworld would be checkered light and dark. "The builders, if they're something like humans, will want day and night because they will want an imitation of their own planet," Niven says.

Solar panels on the immense shadows squares could collect energy to power the structure. Energy could be beamed via laser from the squares to receiver stations along the Ringworld's rim, away from inhabited "land."

Lasers would also come in handy for vaporizing asteroids or comets that might smack into the Ringworld. As a big, thin target, a Ringworld would be devastated by a high-speed impactor. A hole explosively punched through it could let the atmosphere eventually drain out.

Impossible Strength?

Material strength is a potential showstopper for a Ringworld. Because of its bulk, the megastructure would be subjected to mechanical stresses violent enough to break any known physical molecular bonds. "The Ring needs to be superstrong," Sandberg says. "Mere molecular bonds will not do."

For super-strength, the best would bet would be, well, the "strong" force. This force is the grippiest of the four described forces of nature. It has 137 times the strength of electromagnetism, a million times that of the weak force, and duodecillion (10^{39}) times that of puny gravity. Yet it operates only on the femtometer scale of the atomic nucleus. The strong force crams like-charged protons into an atomic nucleus. "The electromagnetic repulsion between the [protons] would love to split them apart, but you have the strong nuclear force gluing them together," Sandberg says.
In our present technological state, we are quite good at manipulating electromagnetism and dealing with gravity. If we could learn to wield the strong force, it would suffice for the structural integrity of a Niven Ring. The strong force is medicated by particles called gluons; if we could rip apart quarks and use

their "glue" beyond the nucleic scale, all sorts of architectural and engineering feats would become possible.

"We have no clue how to control the strong nuclear force," Sandberg says, "but it could be that advanced civilizations know how."

Niven avoided this can of worms in his stories by inventing a magic, milky-gray material called "scrith." He envisioned it being somehow producible by transmutation of elements, via high-tech fusion. Transmutation of elements, such as the predominant hydrogen and helium available within Jupiter and Saturn, would be necessary anyhow for enough (non-scrith) material to build the megastructure.

From Worlds to a Ringworld

As for the actual Ringworld building process, Niven sketched it as follows. The solar system's planets would be dismantled by machines and reformatted into disc-shaped plates. Cables would link these plates and, in time, the plates would be pulled together to form a ring.

Given the miracle materials and advanced element transmutation required for a colossal Ringworld, smaller, other ringlike habitats make far more sense from an engineering perspective. The "Halos" in the eponymous video games, for instance, are about 10,000 miles in diameter.

They could plausibly be made of steel. Bishop Rings, another proposed ring megastructure by a nanotechnologist, Forrest Bishop, would be a "mere" 1,200 miles in diameter and made of ultra-stiff carbon nanotubes. These rings would not encircle a star or planet, but could nestle stably in a Lagrangian point, where the gravitational pull from a planet matches that of the sun.

A ship swoops toward a Halo ring, under construction. The Ark, a construction and control station for Halos, is seen in the bottom of the image.

Finally, the rationale for ever pursuing a Ringworld is questionable in the first place. The civilization's rulers would be placing an awful lot of eggs in one basket. A catastrophic failure somewhere on the Ring, perhaps of a stabilizing thruster, could doom the entire venture, and its trillion of inhabitants. (Niven explores this kind of crisis in *The Ringworld Engineers*.)

Niven himself points out that Ringworlds are really for telling a good story rather than offering a prescription for an Earth whose population has runneth over.

"Even if we go for big stuff, there is no reason to build a Ringworld," Niven says, "when we could build a million [other] things and put them in orbit, rather than in orbit around the sun."

18.2 Dyson Spheres

The most advanced space construction we have conceptualized is for a Dyson Sphere which is a sphere enclosing a star. (Freeman Dyson was the originator) The idea is to harness all of the energy of an entire star.

Here is an article on how that might be done:

How to build a Dyson sphere in five (relatively) easy steps

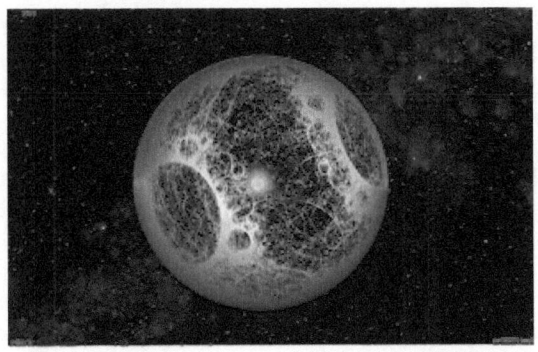

Let's build a Dyson sphere! And why wouldn't we want to?

By enveloping the sun with a massive array of solar panels, humanity would graduate to a Type 2 Kardashev civilization capable of utilizing nearly 100% of the sun's energy output. A Dyson sphere would provide us with more energy than we would ever know what to do with while dramatically increasing our living space.

Given that our resources here on Earth are starting to dwindle, and combined with the problem of increasing demand for more energy and living space, this would seem to a good long-term plan for our species. Implausible you say? Something for our distant descendants to consider?

Think again: We are closer to being able to build a Dyson Sphere than we think. In fact, we could conceivably get going on the project in about 25 to 50 years, with completion of the first phase requiring only a few decades. Yes, really.

Now, before I tell you how we could do such a thing, it's worth doing a quick review of what is meant by a "Dyson sphere".

Dyson Spheres, Swarms, and Bubbles

The Dyson sphere, also referred to as a Dyson shell, is the brainchild of the physicist and astronomer Freeman Dyson. In 1959 he put out a two page paper titled, "Search for Artificial Stellar Sources of Infrared Radiation" in which he described a way for an advanced civilization to utilize all of the energy radiated by their sun. This hypothetical megastructure, as envisaged by Dyson, would be the size of a planetary orbit and consist of a shell of solar collectors (or habitats) around the star. With this model, all (or at least a significant amount) of the energy would hit a receiving surface where it can be used. He speculated that such structures would be the logical consequence of the long-term survival and escalating energy needs of a technological civilization.

Needless to say, the amount of energy that could be extracted in this way is mind-boggling. According to Anders Sandberg, an expert on exploratory engineering, a Dyson sphere in our solar system with a radius of one AU would have a surface area of at least 2.72×10^{17} km2, which is around 600 million times the surface area of the Earth. The sun has an energy output of around 4×10^{26} W, of which most would be available to do useful work.

I should note at this point that a Dyson sphere may not be what you think it is. Science fiction often portrays it as a solid shell enclosing the sun, usually with an inhabitable surface on the inside. Such a structure would be a physical impossibility as the tensile strength would be far too immense and it would be susceptible to severe drift.

Dyson's original proposal simply assumed there would be enough solar collectors around the sun to absorb the starlight, not that they would form a continuous shell. Rather, the shell would consist of independently orbiting structures, around a million kilometers thick and containing more than 1x10⁵ objects.

Consequently, a "Dyson sphere" could consist of solar captors in any number of possible configurations. In a Dyson swarm model, there would be a myriad of solar panels situated in various orbits. It's generally agreed that this would be the best approach. Another plausible idea is that of the Dyson bubble in which solar sails, as well as solar panels, would be put into place and balanced by gravity and the solar wind pushing against it.

For the purposes of this discussion, I'm going to propose that we build a Dyson swarm (sometimes referred to as a type I Dyson sphere), which will consist of a large number of independent constructs orbiting in a dense formation around the sun. The advantage of this approach is that such a structure could be built incrementally. Moreover, various forms of wireless energy transfer could be used to transmit energy between its components and the Earth.

Megascale construction

So, how would we go about the largest construction project ever undertaken by humanity?

As noted, a Dyson swarm can be built gradually. And in fact, this is the approach we should take. The primary challenges of this approach, however, is that we will need advanced materials (which we still do not possess, but will likely develop in the coming decades thanks to nanotechnology), and autonomous robots to mine for materials and build the panels in space.

Now, assuming that we will be able to overcome these challenges in the next half-decade or so—which is not too implausible— how could we start the construction of a Dyson sphere?

Oxford University physicist Stuart Armstrong has devised a rather ingenious and startling simple plan for doing so—one which he claims is almost within humanity's collective skill-set. Armstrong's plan sees five primary stages of construction, which when used in a cyclical manner, would result in increasingly efficient, and even exponentially growing, construction rates such that the entire project could be completed within a few decades.

Broken down into five basic steps, the construction cycle looks like this:

1. Get energy
2. Mine Mercury
3. Get materials into orbit
4. Make solar collectors
5. Extract energy

The idea is to build the entire swarm in iterative steps and not all at once. We would only need to build a small section of the Dyson sphere to provide the energy requirements for the rest of the project. Thus, construction efficiency will increase over time as the project progresses. "We could do it now," says Armstrong. It's just a question of materials and automation.

And yes, you read that right: we're going to have to mine materials from Mercury. Actually, we'll likely have to take the whole planet apart. The Dyson sphere will require a horrendous amount of material—so much so, in fact, that, should we want to completely envelope the sun, we are going to have to disassemble not just Mercury, but Venus, some of the outer planets, and any nearby asteroids as well.

Why Mercury first? According to Armstrong, we need a convenient source of material close to the sun. Moreover, it has a good base of elements for our needs. Mercury has a mass of 3.3×10^{23} kg. Slightly more than half of its mass is usable, namely iron and oxygen, which can be used as a reasonable construction material (i.e. hematite).

So, the useful mass of Mercury is 1.7×10^{23} kg, which, once mined, transported into space, and converted into solar captors, would create a total surface area of

245g/m2. This Phase 1 swarm would be placed in orbit around Mercury and would provide a reasonable amount of reflective surface area for energy extraction.

There are five fundamental, but fairly conservative, assumptions that Armstrong relies upon for this plan. First, he assumes it will take ten years to process and position the extracted material. Second, that 51.9% of Mercury's mass is in fact usable. Third, that there will be 1/10 efficiency for moving material off planet (with the remainder going into breaking chemical bonds and mining). Fourth, that we'll get about 1/3 efficiency out of the solar panels. And lastly, that the first section of the Dyson sphere will consist of a modest 1 km2 surface area.

And here's where it gets interesting: Construction efficiency will at this point start to improve at an exponential rate.

Consequently, Armstrong suggests that we break the project down into what he calls "ten year surges." Basically, we should take the first ten years to build the first array, and then, using the energy from that initial swarm, fuel the rest of the project. Using such a schema, Mercury could be completely dismantled in about four ten-year cycles. In other words, we could create a Dyson swarm that consists of more than half of the mass of Mercury in forty years! And should we wish to continue, if would only take about a year to disassemble Venus.

And assuming we go all the way and envelope the entire sun, we would eventually have access to 3.8×10^{26} Watts of energy.
Dysonian existence

Once Phase 1 construction is complete (i.e. the Mercury phase), we could use this energy for other purposes, like megascale supercomputing, building mass drivers for interstellar exploration, or for continuing to build and maintain the Dyson sphere.

Interestingly, Armstrong would seem to suggest that this might be enough energy to serve us. But other thinkers, like Sandberg, suggest that we should keep going. But in order for us to do so we would have to deconstruct more planets. Sandberg contends that both the inner and outer solar system contains enough usable material for various forms of Dyson spheres with a complete 1 AU radius (which would be around 42 kg/m2 of the sphere). Clearly, should we wish to truly attain Kardashev II status, this would be the way to go.

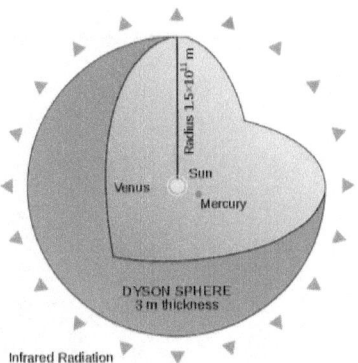

And why go all the way? Well, it's very possible that our appetite for computational power will become quite insatiable. It's hard to predict what a post-Singularity or post-biological civilization would do with so much computation power. Some ideas include ancestor simulations, or even creating virtual universes within universes. In addition, an advanced civilization may simply want to create as many positive individual experiences as possible (a kind of utilitarian imperative). Regardless, digital existence appears to be in our future, so computation will eventually become our most valuable and sought after resource.

That said, whether we build a small array or one that envelopes the entire sun, it seems clear that the idea of constructing a Dyson sphere should no longer be relegated to science fiction or our dreams of the deep future. Like other speculative projects, like the space elevator or terraforming Mars, we should seriously consider putting this alongside our other near-term plans for space exploration and work.

And given the progressively worsening condition of Earth and our ever-growing demand for living space and resources, we may have no other choice.

19.0 Long Term Planning

I hope you get some of my enthusiasm from this book about the long term value and excitement from building, living, and working in space structures.

Again, the tools to build large space habitats are a matter of technology, not physics. These tools will take a lot more research and development and could take anywhere from ten years to one hundred years or more to have them ready to start these ambitious projects.

Having coordination of these efforts with government and industrial planning would help decrease the time significantly before we are ready to build the large space habitats.

What we need to do is provide incentives and directions to minimize the time before these technologies come to market. My suggestions include:

1) Develop a private enterprise and government committee to regularly check status and provide helpful guides for required technologies to maturity to be available this century for usage in building space habitats.

2) In conjunction with #1 develop series of X Type Prize commitments to provide funding for reaching certain milestones. One milestone might be to provide a prize for the first production of a construction robot which could replicate itself. It could even be a series of prizes with gradually more complex capabilities resulting in a final replicating construction robot.

3) Government providing funding research to many of these required technologies through agencies like the Defense Department's DARPA. This

funding would help these technologies to mature sooner rather than later.

4) Industry and Government committees to develop a schedule or roadmap for building the first large space habitat and get commitments from all to help meet that timetable. Members would include not only the committee, but technology suppliers too.

Also, here are the milestones of the National Space Society website for making progress on space settlements:

Part II: MILESTONES TO ALL DESTINATIONS
GENERAL BARRIERS TO SPACE SETTLEMENT
- Psychological
- Social
- Economic

MILESTONE 1: Continuous Occupancy in Low Earth Orbit.
Construction of continuously occupied structures in Low Earth Orbit (LEO).

MILESTONE 2: Higher Commercial Launch Rates and Lower Cost to Orbit.

The emergence of a sufficiently large launch market, with more efficient and reliable vehicles with faster turnaround times, or technical and operational improvements such as re-usable vehicles, or both, significantly lowering the cost of access to space. Both higher launch rates and lower vehicle and operational costs will be required.

- Flight Test Demonstrations
- Government Contracting Practices
- Progress in Launch Technology
- Space Tourism

- Commercial Facilities in Orbit
- Space Solar Power
- Other Commercial Space Applications
- Governmental Policies

MILESTONE 3: An Integrated Cislunar Space Transportation System.

In addition to Earth-to-orbit launch systems, the creation of transportation systems and infrastructure in "cislunar space," i.e., the space between the Earth and the Moon, resulting in regular commerce in cislunar space.

MILESTONE 4: Legal Protection of Property Rights.
Legal protection of property rights enacted to provide prospective off-Earth investors and settlers with the security to take financial risks.

MILESTONE 5: Land Grants or Other Economic Incentives.
Economic incentives, such as "land grants," to encourage private investment in off-Earth settlements.

MILESTONE 6: Technology for Adequate Self-Sufficiency.
People leaving Earth with the technology and tools needed to settle, survive and prosper without needing constant resupply from Earth.
- Enabling Technologies
- Precursor Missions

Part III. UTILIZATION OF SPACE TECHNOLOGY AND RESOURCES

MILESTONE 7: Applications of Space Technology on and for Earth.

The technologies and techniques developed on the road to space settlement applied widely and benefiting all on Earth.

In General: Direct Benefits
In General: Indirect Benefits
Later: Extra-Terrestrial Raw Materials
Later: Sunlight from Orbit

<u>MILESTONE 8</u>: Space Solar Power System (SSP). Establishment of an operational space-based solar power system transmitting the Sun's energy to Earth.

MILESTONE 9: A Workable Asteroid Protection System.

A system capable of detecting and defending against Earth-approaching asteroids or comets built and standing by to launch on short notice.

Part IV. TO THE MOON

PARTICULAR BARRIERS:

- Political and Psychological
- Goal Definition
- Biological
- Uniquely Lunar

MILESTONE 10: Robotic Confirmation of Lunar Resources.
Satellites orbiting the Moon and possibly robotic landers determining the nature and extent of lunar ice and volatile deposits and providing the information necessary to guide the choice of the best sites for a lunar outpost.
- Water and Lunar Volatiles
- Sites for Lunar Outposts

MILESTONE 11: A Lunar Research Facility.
A lunar research facility established to study human habitation, test various equipment and techniques, and conduct lunar investigations.

MILESTONE 12: A Government / Industry Lunar Base.

The initial research facility evolving into a permanently occupied, ever-expanding lunar base, or such a base created at another site using what has been learned from the initial facility, and increasingly performing commercial functions.

- Noncommercial Functions
- Commercial Functions

MILESTONE 13: A True Lunar Settlement. The lunar base evolving into a permanent settlement, increasingly self-sufficient and increasingly focused on commercial activities.

Part V. TO MARS

PARTICULAR BARRIERS:

- Psychological and Political
- Goal Definition
- Biological
- Uniquely Martian

MILESTONE 14: Robotic Exploration of Mars for Local (In Situ) Resources.

Satellites orbiting Mars and robotic landers determining the nature and extent of Martian resources, especially ice, guiding the choice of the best sites for follow-on human missions.

- Scientific Knowledge
- Data for Human Exploration and Outposts

MILESTONE 15: Creation of a Logistics System for Transporting Humans and Cargo to the Martian Surface.

An integrated sustainable system designed and built for transporting humans and cargo from space to the Martian surface, maintaining the crew on the surface, and returning crew and payload safely back to Earth.

- Earth to Mars Transit Transportation Systems
- Earth Orbit to Mars Orbit on a "Cycler"
- Mars Landing Systems
- Earth Return Systems
- Martian "Ferries"
- Orbital Propellant Depot
- Habitation Systems

MILESTONE 16: A Continuously Occupied Multi-Purpose Base.

Following initial crewed missions to identify a suitable location and the particular infrastructure and equipment needed there, establishment of a continuously occupied multi-purpose Mars base.

MILESTONE 17: A True Martian Settlement.
The Martian base evolving into a permanent settlement, increasingly self-sufficient and increasingly focused on commercial activities.

Part VI. TO THE ASTEROIDS

MILESTONE 18: Exploration, Utilization and Settlement of Asteroids.

After robotic identification of suitable asteroids, robotic and human crews following to establish mining bases and habitats for transients, and, eventually, carving out and building permanent human settlements.

Part VII. TO ORBITAL SPACE SETTLEMENTS

<u>MILESTONE 19</u>: Construction of Orbital Space Settlements.
Orbital "cities in space" built from asteroid or lunar materials.

20.0 Summary

I hope you enjoyed reading this book on living and working in Outer Space. I tried to make it fun as well as providing useful information and recommendations to help this process develop.

Aside from spiritual development, I think this is the most glorious future humanity can look forward to by becoming a cosmic species.

Getting humanity living and working in space offers these long term benefits:

1) Humanity will no longer be subject to extinction because of a nuclear war or other Earth cataclysm like a huge asteroid hitting it.

2) Settlement of space and other planets and structures in space will provide the same renaissance of world cultures that the exploration and settlement of Earth did centuries ago—only to an unlimited degree.

3) We will be building the infrastructure we will need for a true Solar System civilization and the means to travel to the nearest stars.

I look forward to living long enough so I can visit some of these Space Habitats. That would be really fun.

Martin Ettington

Revised March 2023

21.0 Bibliography

1. O'Neill, Gerard. *The High Frontier: Human Colonies in Space.* s.l. : William Morrow & Company, 1977.

2. Wiki. The High Frontier: Human Colonies in Space. *Wikipedia.* [Online] https://en.wikipedia.org/wiki/The_High_Frontier:_Human_Colonies_in_Space.

3. 3D Printing of Large Structures. [Online] [Cited: 6 6, 2016.] http://3dprintingindustry.com/news/how-viable-is-3d-printing-for-building-large-structures-75032/.

4. Extracting Graphine from Graphite. [Online] [Cited: 6 6, 2016.] http://science.wonderhowto.com/how-to/make-graphene-sheets-from-graphite-flakes-and-cellophane-tape-402113/.

5. Plan to Turnig Asteroids into Spaceships. [Online] [Cited: 6 7, 2016.] http://www.space.com/33079-turning-asteroids-into-spaceships-made-in-space.html.

6. Hale, Edward Everett. *The Brick Moon.* 1869.

7. Brand, Stewart. *Space Colonies.* s.l. : Penguin, 1977.

8. Science, NASA. https://science.nasa.gov/science-news/science-at-nasa/2001/ast21mar_1. *http://science.nasa.gov.* [Online] 2017.

9. Research, Space Station. https://www.nasa.gov/mission_pages/station/research/news/sabatier.html. *https://www.nasa.gov.* [Online] 2017.

10. SpaceX. *spacex.com.* [Online] 2017.

11. Vogt, A.E. Van. *The Voyage of the Space Beagle* . 1950.

12. Various. http://news.cornell.edu/stories/2005/05/researchers-build-robot-can-reproduce. *http://news.cornell.edu.* [Online] 2005.

13. http://infinitysedge.weebly.com/space-habitats.html. *http://infinitysedge.weebly.com.* [Online] Twin O'Neil Space Habitats, 2017.

14. https://en.wikipedia.org/wiki/Asteroid_mining. *https://en.wikipedia.org.* [Online] Asteroid Mining, 2017.

15. https://www.mars-one.com/mission/mars-one-astronauts. *https://www.mars-one.com.* [Online] Mars One, 2017.

16. http://www.esa.int/Our_Activities/Space_Engineering_Technology/Onboard_Computer_and_Data_Handling/Architectures_of_Onboard_Data_Systems. *http://www.esa.int.* [Online] ISS Data ARchitecture, 2017.

17. https://www.nasa.gov/mission_pages/tdm/sep/index.html . *https://www.nasa.gov.* [Online] Solar Electric Propulsion, 2017.

18. https://www.nasa.gov/centers/glenn/about/fs21grc.html. *https://www.nasa.gov.* [Online] Ion Propulsion, 2017.

19. http://emdrive.com/. *http://emdrive.com/.* [Online] The EM DRive, 2017.

20. https://en.wikipedia.org/wiki/ISS_ECLSS. *https://en.wikipedia.org.* [Online] ECLSS on ISS, 2017.

21. https://www.nasa.gov/pdf/167129main_Systems.pdf. *https://www.nasa.gov.* [Online] ISS Systems, 2017.

22. http://www.spacefuture.com/tourism/hotels.shtml. *http://www.spacefuture.com.* [Online] Space Hotels, 2017.

23. http://www.sentientdevelopments.com/2012/03/how-to-build-dyson-sphere-in-five.html. *http://www.sentientdevelopments.com.* [Online] Building a Dyson Sphere, 2017.

24. https://en.wikipedia.org/wiki/Mars_One#Mission_proposals. *https://en.wikipedia.org.* [Online] Mars One, 2017.

25. https://www.nasa.gov/feature/nasa-plant-researchers-explore-question-of-deep-space-food-crops. *https://www.nasa.gov.* [Online] NASA Plant Production, 2017.

26. http://www.nss.org/settlement/library.html. *http://www.nss.org.* [Online] National Space Society Online Library, 2017.

27. http://www.nss.org/settlement/space/#intro. *http://www.nss.org.* [Online] NSS Orbital Space Settlements Intro, 2017.

28. http://www.lunarpedia.org/index.php?title=Mass_Drivers. *http://www.lunarpedia.org.* [Online] Lunar Mass Drivers, 2017.

29. http://www.nss.org/settlement/DistantSuns/distantsuns_chap08.html. *http://www.nss.org.* [Online] Paths to Commerce, 2017.

30. Cook, Nick. *The Hunt for Zero Point.* s.l. : Broadway Books, 2001.

31. https://www.nasa.gov/press-release/nasa-darpa-will-test-nuclear-engine-for-future-mars-missions. *New Nuclear Propulsion Rocket.* [Online] 2023.

22.0 Index

3D Printing, 139
3D Systems, 139
A. E. van Vogt, 94
anti-gravity, 25
Armadillo Aerospace, 60
Artemus Moon Landers, 131
Asteroid Homes, 207
Asteroid Mining, 137, 158
   Heating, 160
   Magnetic rakes, 160
   Shaft mining, 160
   Surface mining, 160
Atmosphere, 50
Bases on the Moon, 77
Batteries, 40
Blue Origin, 59, 123
Breathable Atmosphere, 34
Canada, 65
Cargo Orbit Transfer Vehicles, 58
Cargo rockets, 58
Challenger Explosion, 2
circular mass drive, 156
Citizens in Space, 60
Communication protocols, 44
counter rotating O'Neil Habitats, 16
Data Architecture/Communications, 42
diamond nanothreads, 162
Dream Chaser, 61, 127
Dyson Sphere, 223
EADS Astrium, 61
Earth Protector-The Psychic Soldier Series Book 4, 207
ECLSS, 35
ESA, 66
ExOne, 139
Food Production, 45
Fresh Water, 30
Fusion Reactors, 108
Giant Chinese Space Station, 73
Government Vs Free Enterprise, 56
Guidance and Control, 51
Habitat Cylinders, 163
Heat and Cold, 28
Hotels in Orbit, 81
HP Inc., 140
Hudson's Bay Company, 55
Inflatable structures, 92
interior of a Bernal Sphere, 20
International Space Station, 5
Ion Propulsion, 112
Island One, 19
ISS Electrical Power Distribution, 41
Jaxa, 65
Kalpana One, 20
Konstantin Tsiolkovsky, 13, 162
Lagrange 5 point, 164
Larry Niven, 216
Lunar Gateway Partners, 65

Mars One, 87
Mars One project, 87
Mars Settlement, 63
mass accelerator, 156
Materials for Construction, 153
microscopic spheres, 61
Mining the Moon, 154
Nuclear Propulsion, 103
Nuclear Propulsion Contract, 104
Nuclear Pulse Propulsion, 107
O'Neil Cylinders, 14
Orbital Reef, 69
Paths to Commerce, 57
Personnel Orbit Transfer Vehicles, 58
Planetary Resources, 137
Power Systems, 39
powersats, 58
Pressure Containment, 36
Professor Dr. Gerard K. O'Neill, 9
Project Orion, 107
Propulsion Systems, 109
protein crystal growth, 63
Radiation Protection, 27
Reduced launch costs, 92
Relatvity Space, 141
Remote Terminal Unit, 43
Ringworld, 216
Robert Heinlein, 77
Robotics, 94
Roscosmos, 66
self replicating robots, 94, 95
Shackleton Crater, 78
Solar Array Wings, 39
Solar Smelting, 101
Space Colonies, 10
Space Design Studies, 21
Space Elevators, 161
Space Expedition Corporation, 60
Space Exploration Technologies Corp, 115
Space Infrastructure Development, 91
Space Manufacturing, 61
Space Settlements-A Design Study, 21
Space Tourism, 58
Space X-Rocket History, 115
Spacecraft Management Unit, 42
SpaceShipOne, 59
SpaceX Dragon capsule, 56
SpaceX Starship, 119
Stanford Torus inside view, 18
Stanford Torus outside view, 18
Starlab, 71
Stewart Brand, 10
The Bernal Sphere, 19
The Brick Moon, 9
The Deep Space Gateway, 66

The EM Drive, 113
The Fountains of Paradise, 161
The High Frontier, 9, 14
The Hunt for Zero Point, 25
The Moon is a Harsh Mistress, 77
The Stanford Torus, 17
The Voyage of the Space Beagle, 94
Three Dimensional Printing, 100
Tiangong Space Station, 67
TM/TC, 43
Virgin Galactic, 59
Waste Elimination, 38
Wernher von Braun, 17
XCOR Aerospace, 59

www.ingramcontent.com/pod-product-compliance
Lightning Source LLC
Chambersburg PA
CBHW021811170526
45157CB00007B/2539